COORDINATED SCIENCE

SECOND EDITION

Chemistry

Mary Jones, Geoff Jones & David Acaster

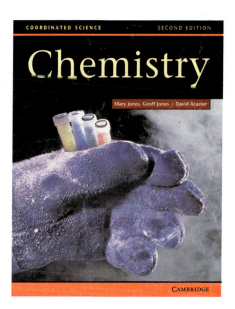

CAMBRIDGE
UNIVERSITY PRESS

PUBLISHED BY THE PRESS SYNDICATE OF THE UNIVERSITY OF CAMBRIDGE
The Pitt Building, Trumpington Street, Cambridge CB2 1RP, United Kingdom

CAMBRIDGE UNIVERSITY PRESS
The Edinburgh Building, Cambridge CB2 2RU, United Kingdom
40 West 20th Street, New York, NY 10011-4211, USA
10 Stamford Road, Oakleigh, Melbourne 3166, Australia

© Cambridge University Press 1997
Artwork © Geoff Jones (except figure 71.2)

First published 1993
Second edition 1997

Printed in the United Kingdom at the University Press, Cambridge

A catalogue record for this book is available from the British Library

ISBN 0 521 59983 0

Prepared for the publishers by Stenton Associates

Cover photograph: Sample tubes just removed from cold storage in
liquid nitrogen, Geoff Tompkinson/Science Photo Library

CONTENTS

How to use this book

HOW TO USE THIS BOOK

The material is organised into main **subject areas**. Each has its own colour key which is shown on the contents list on page 3 and on the contents lists appearing at the beginning of each subject area.

Core text is material suitable for students of a wide ability range. However, not all of the core text will be suitable for all syllabuses and teachers should be aware of this.

70 DISPLACEMENT REACTIONS

The best way to compare metal reactivity is by using displacement reactions.

Reactive metals can displace less reactive metals from their compounds

When magnesium is heated in steam, hydrogen is formed.

magnesium + steam → magnesium oxide + hydrogen

$$Mg(s) + H_2O(g) \rightarrow MgO(s) + H_2(g)$$

We say that the magnesium has **displaced** the hydrogen. The magnesium has taken the place of the hydrogen in the oxide compound.

You will remember that a reactive element forms compounds more readily than an unreactive one. In this reaction, magnesium is becoming part of a compound, magnesium oxide. The displaced hydrogen is leaving its compound, water. This shows that magnesium is more reactive than hydrogen.

Displacement reactions can also occur between metals. For example:

copper (II) oxide + magnesium → magnesium oxide + copper

$$CuO(s) + Mg(s) \rightarrow MgO(s) + Cu(s)$$

This is a very vigorous reaction! The copper is displaced by the magnesium. This confirms the results of the reactions on the previous two pages – magnesium is more reactive than copper. Just to check, we can heat magnesium oxide with copper. Nothing happens. Magnesium cannot be displaced by copper.

Fig. 70.1 Magnesium reacting with copper (II) oxide.

Displacement reactions tell you which one of two metals is the more reactive

Displacement reactions give much more positive evidence about reactivity than comparing the reactions of metals with water or acid. If metal A displaces metal B, then metal A is more reactive than metal B. If metal B displaces metal A, then metal B is the more reactive of the two.

We can now, at last, sort out the relative reactivities of lead and copper! If lead oxide is heated with copper, nothing happens. If copper (II) oxide is heated with lead, lead oxide is produced. So lead displaces copper from the copper compound. Lead is more reactive than copper.

One famous displacement reaction demonstrates that aluminium is more reactive than iron. It is called the 'Thermit' reaction. Aluminium powder is mixed with iron (III) oxide powder. The mixture is heated to start the reaction. Once started, the reaction is very rapid. A large amount of heat is given off. The products are aluminium oxide and some very hot iron. This reaction has been used to produce molten iron, to weld lengths of railway line together.

Fig. 70.2 The Thermit reaction in use. Molten iron has formed and is flowing down to fill the gap between two rails at the bottom.

Solutions of salts can be used to perform displacement reactions

Displacement reactions like the Thermit reaction can be very exciting, but they are also dangerous. The Thermit reaction produces molten iron, which can do a lot of damage. A safer way of performing displacement reactions in the laboratory is by using solutions of metal salts, instead of the metal oxide. A piece of a pure metal is put into a solution of the metal salt of a different metal. If the pure metal is more reactive than the metal in the salt, then the metal in the salt is displaced. The metal in the salt appears in the liquid as a pure metal, often as whisker-like growths on the surface of the pure metal you added originally.

For example, copper is less reactive than magnesium. If a piece of magnesium is put into a beaker containing some copper (II) sulphate solution, the magnesium displaces the copper from its compound. Some of the displaced copper sticks to the magnesium, while some falls to the bottom of the beaker. The magnesium ribbon slowly disappears, as the magnesium takes the place of copper in the solution – magnesium sulphate is formed.

$$magnesium + copper (II) sulphate \rightarrow magnesium sulphate + copper$$

$$Mg(s) + CuSO_4(aq) \rightarrow MgSO_4(aq) + Cu(s)$$

Of course, if you try putting a piece of copper into some magnesium sulphate solution, nothing will happen. Copper cannot displace magnesium.

Fig. 70.3b When the magnesium is taken out after a minute, copper is clearly visible.

Fig. 70.3a Magnesium is added to copper (II) sulphate solution.

Displacement reactions and redox
In all displacement reactions, oxidation and reduction takes place. Metallic elements lose electrons, and form compounds. They are oxidised. Other metals, in compounds, gain electrons, and become pure elements. They are reduced.

For example:

$$copper (II) oxide + magnesium \longrightarrow magnesium oxide + copper$$
$$Cu^{2+}O^{2-}(s) + Mg(s) \longrightarrow Mg^{2+}O^{2-}(s) + Cu(s)$$

Half equations:

$Cu^{2+} + 2e^- \longrightarrow Cu$ Copper is reduced
$Mg - 2e^- \longrightarrow Mg^{2+}$ Magnesium is oxidised

Another example:

$$zinc sulphate + magnesium \longrightarrow magnesium sulphate + zinc$$
$$Zn^{2+}SO_4^{2-}(aq) + Mg(s) \longrightarrow Mg^{2+}SO_4^{2-}(aq) + Zn(s)$$

Half equations:

$Zn^{2+} + 2e^- \longrightarrow Zn$ Zinc is reduced
$Mg - 2e^- \longrightarrow Mg^{2+}$ Magnesium is oxidised

— EXTENSION —

Questions

1. When nickel is heated with copper (II) oxide, nickel oxide and copper are formed.
   a. Which of the two metals is more reactive?
   b. What other experiments could you do to check on this?
2. Which of the following mixtures would you expect to react when heated? Give products for those which react.
   a. zinc and copper (II) oxide
   b. zinc and lithium oxide
   c. copper and iron (III) oxide

   d. magnesium and calcium oxide
   e. lithium and calcium oxide
   f. iron and lead (II) oxide
3. a. You are given samples of manganese and iron, and samples of their soluble salts manganese (II) chloride and iron (II) chloride. How could you find out which metal is more reactive?
   b. Manganese is more reactive than iron. What would you see happening if you performed the experiment you suggest in your answer to part a?

4. Why would you not try to prove that lithium is more reactive than magnesium by putting lithium into magnesium sulphate solution?

5. a. Write a word equation for the Thermit reaction.
   b. Write a balanced equation for this reaction.
   c. Write half equations for the metals involved, and identify the metal which is oxidised and the metal which is reduced.

— EXTENSION —

INVESTIGATION 70.1

The reactivity of silver

Silver is an unreactive metal. This is why it is used for jewellery. It does not react with your skin. But is silver more, or less, reactive than copper? Find out by performing these displacement reactions.

1 Label two test tubes 'A' and 'B'. Into tube A, put a 5 cm length of thick, clean copper wire. Cover the wire with silver nitrate solution. Do not shake the tube. Leave it to stand for 30 min. Observe the contents of the tube, and record your observations.

2 Into tube B, put a 5 cm length of clean silver wire. Cover the wire with copper (II) sulphate solution. Do not shake the tube. Leave it to stand for 30 min. Observe the contents of the tube, and record your observations.

Questions

1 In which tube was there evidence of a reaction? What was this evidence?
2 Which metal displaced the other?
3 Which metal is more reactive?
4 Write a word equation for the reaction which takes place.
5 Write a balanced equation for the reaction which takes place.

167

ELEMENTS AND COMPOUNDS

1 ATOMS

All material is made from tiny particles, called atoms.

All substances are made of atoms

Everything that you see around you is made out of tiny particles called **atoms**. This page, the ink on it, you and your chair are all made of atoms. Atoms are remarkably small. The head of a pin contains about 60 000 000 000 000 000 000 atoms. It is difficult to imagine anything quite so small as an atom. The ancient Greeks believed that nothing smaller than an atom could exist, so they gave them the name 'atomos', meaning 'indivisible'. Atoms sometimes exist singly, and sometimes in groups. These groups of atoms are also known as **molecules**.

The way substances behave suggests that they are made of tiny particles

Atoms are far too small to be seen. Yet we know they must be there, because of the way that substances behave.

Crystals of many materials have regular shapes. The crystals of a particular substance always have the same shape. One explanation for this is that crystals are built up of tiny particles, put together in a regular way. These tiny particles are atoms.

Fig. 1.1 Sodium chloride (salt) crystals. The actual size of the biggest crystal in this enlarged photograph is about 0.3 mm across.

A crystal such as the one in the photograph contains millions upon millions of atoms. If a small piece is chipped away from a salt crystal, for example, the piece is still salt. Even the very smallest, microscopic piece you can chip away is still salt.

Fig. 1.2 A salt crystal. Sodium chloride crystals are cube-shaped. This is because they are made of particles which are arranged in a regular, cubic pattern. The diagram shows the particles in a very, very tiny piece of sodium chloride.

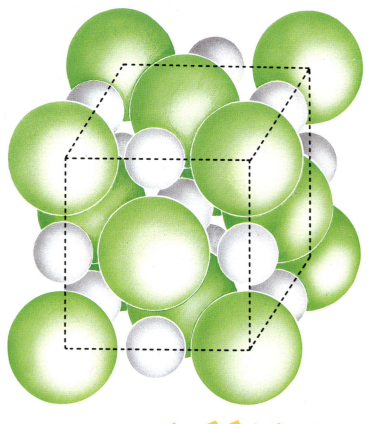

DID YOU KNOW?

If you divided a single drop of water so that everyone in the world had an equal share, everyone would still get about one million million molecules.

Brownian motion is evidence for the existence of particles

In 1827, an Oxford scientist, Robert Brown, was studying pollen grains through a microscope. The pollen grains were in liquid, and Brown was surprised to see that the movement was jerky. Other scientists became interested in Brown's observations. Many suggestions were made to try to explain them. Some people thought that the pollen might be moving of its own accord. Others suggested that it was being knocked around by something invisible in the liquid in which it was floating.

You can see this effect with the apparatus in Figure 1.4, using smoke grains instead of pollen grains.

No-one could possibly imagine that smoke grains are alive. The jerky movement can best be explained by imagining tiny air molecules around the smoke grains, bumping into them. This knocks the smoke grains around in jerky movements. The air molecules are much too small to be seen. They must be moving around very quickly.

This jerky movement of the pollen grains or smoke grains is called **Brownian motion**, after Robert Brown. It is strong evidence that all substances are made up of molecules, in constant motion, and much too small to be seen.

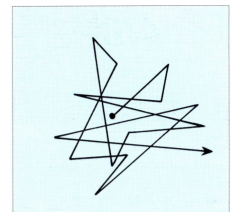

Fig. 1.3 Brownian motion. Small particles like smoke particles seem to move randomly, constantly changing direction.

INVESTIGATION 1.1

Using a smoke cell to see Brownian motion

A smoke cell is a small glass container in which you can trap smoke grains, and then watch them through a microscope.

1 Make sure that you understand the smoke cell apparatus. The smoke cell itself (see Figure 1.4) should be clean, because light must be able to shine through the sides of it. The light comes from a small bulb. It is focused onto the smoke cell through a cylindrical glass lens.

2 Set up a microscope.

3 Now trap some smoke in the smoke cell, in the following way. Set light to one end of a waxed paper straw. If you hold the burning straw at an angle, you can make the smoke pour out of the lower end and into the smoke cell. Quickly place a cover slip over the smoke in the smoke cell.

4 Put the smoke cell apparatus on the stage of the microscope. Switch on the light. Focus on the contents of the smoke cell.

You should be able to see small specks of light dancing around. These are the smoke grains. They look bright because the light reflects off them.

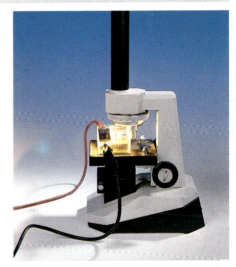

Fig. 1.5 Using a smoke cell

Fig. 1.4 The smoke cell apparatus

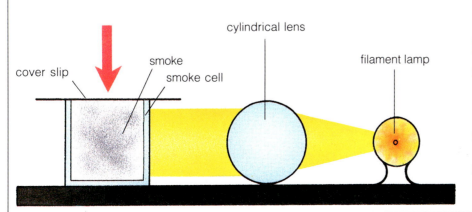

look down through microscope

cylindrical lens

filament lamp

cover slip

smoke

smoke cell

Questions

1 a What can you see in the smoke cell?
b What is in the cell which cannot be seen?
2 Why are the smoke grains dancing around?
3 What is the name for this movement?

2 KINETIC THEORY

All material is made from small moving particles called atoms. Materials can be solid, liquid or gaseous, depending on the arrangement and freedom of movement of these particles.

There are three states of matter

All material is made from tiny particles. These particles are constantly moving. The kinetic theory uses this idea of tiny moving particles to explain the different forms that material can take. 'Kinetic' means 'to do with movement'.

The three states of matter are solid, liquid and gas.

The first state - solid

In this state, matter tends to keep its shape. If it is squashed or stretched enough, it will change shape slightly. Usually, any change in volume is too small to be noticed. The particles are not moving around, although they are vibrating very slightly. Normally they vibrate about fixed positions. If a solid is heated the particles start to vibrate more.

Fig. 2.1a

The second state - liquid

In this state matter will flow. It will take up the shape of any container it is put in. The liquid normally fills a container from the bottom up. It has a fixed volume. If the liquid is squeezed it will change shape, but the volume hardly changes at all. The particles, like those in a solid, are vibrating. However, in a liquid the particles are free to move around each other. If a liquid is heated, the particles move faster.

Fig. 2.1b

The third state - gas

In this state matter will take up the shape of a container and fill it. The volume of the gas depends on the size of its container. If the gas is squashed it will change both volume and shape. The particles are free to move around, and do not often meet each other. The particles whizz around very quickly. If heated they move even faster.

Fig. 2.1c

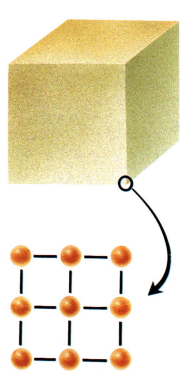

The particles are fixed in position. The forces between particles are strong. The particles cannot move past each other. They are close together.

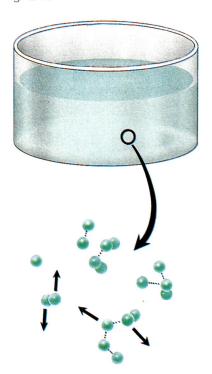

The particles can move past each other. They are joined together in small groups. They are not as close together as in a solid. The forces between the particles are not so strong.

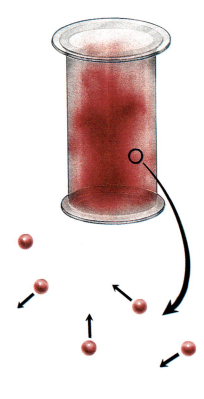

There are hardly any forces between the particles. They are a long way apart. The particles are moving quickly and so they spread out. If squashed, they move closer together.

A model for the kinetic theory

A platform is attached to an electric motor. The motor vibrates the platform up and down. The speed of the vibrations can be controlled.

Fig. 2.2

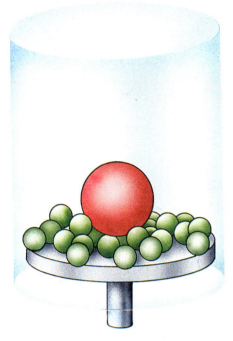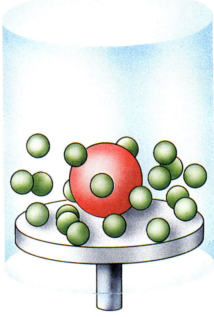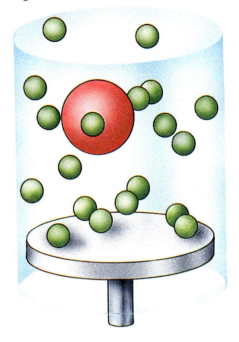

When the platform is still the small balls are like the particles in a solid. They are in fixed positions. If a larger ball is placed on the small ones, it sits on top of them as it would on a solid.

If the platform is made to vibrate gently, the small balls vibrate. But they still stay in the same place. This is what happens when a solid warms up. The particles vibrate, but stay in the same place.

If the platform vibrates faster, the balls start to move faster. It is as though the solid is being heated. When the balls begin to bounce above the level of the platform they are behaving like the particles in a liquid. Now the larger ball is surrounded by the smaller ones. It is as though it has 'sunk' into the liquid.

DID YOU KNOW?

If a gas is heated to a very high temperature, the molecules and then the atoms break apart. At tens of thousands of degrees Celsius, a plasma is formed. The Sun is a plasma. A fluorescent tube contains a plasma.

If the platform vibrates very fast, the balls fly around the whole container. They bounce off the walls. Now they are behaving like the particles in a gas. The large ball is knocked about by the small balls. It is behaving like the smoke grains in the smoke cell experiment.

In this model the average speed of the balls depends on the speed of vibration of the motor. In a substance, the average speed of the particles depends on the temperature. The temperature of a material indicates how quickly the particles are moving.

Fig. 2.3 The Sun is a ball of plasma. The red and blue patches are sunspots.

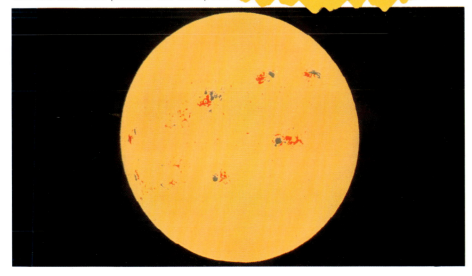

Questions

1 a Name the three states of matter.
  b Describe the arrangement and behaviour of particles in each of the three states.
  c For each of these three states, give one example of a substance which is normally in this state at room temperature.

3 DIFFUSION

Diffusion is the spreading out of particles. It provides further evidence for the kinetic theory.

Moving particles spread around

A smell will slowly spread across a room. We can explain this by imagining that the smell is made of moving particles of a smelly substance. The molecules move around, filling the room. This movement is called **diffusion**. Diffusion can be defined as *the movement of particles from a place where there is a high concentration of them, to a place where there is a lower concentration of them*. Diffusion tends to spread the particles out evenly.

You can watch diffusion happening if you use a substance, such as potassium permanganate, which has coloured particles. If a crystal of potassium permanganate is carefully dropped into a beaker of water, it dissolves. The particles in the crystal separate from each other and slowly move through the water. The colour only spreads slowly. The particles keep bumping into the water molecules. They do not travel in straight lines. Eventually the colour fills the whole beaker, but this takes a very long time.

Diffusion happens faster in gases than in liquids

In a gas the molecules are not so close together as in a liquid. If a coloured gas is mixed with a colourless one, the colour spreads. This happens faster than in a liquid, because fewer particles get in the way. You can watch this happening with bromine, as it diffuses through air. Bromine is a brown gas. It covers about 2 cm in 100 s. If it diffuses into a vacuum, it goes even faster, because there are no particles to get in the way. It then travels 20 km in 100 s!

Fig. 3.1 Potassium permanganate diffusing in water. A crystal has just been dropped into the gas jar on the left. The potassium permanganate has been diffusing for about 30 minutes in the centre jar, and for 24 hours in the jar on the right.

Fig. 3.2 An experiment showing the diffusion of bromine gas. On the left, the two gas jars are separated by a glass lid. The lower one contains bromine, and the upper one contains air. On the right, the lid has been removed. The bromine and air diffuse into one another.

INVESTIGATION 3.1

How quickly do scent particles move?

If a bottle of perfume is opened, the smell spreads across a room. Design an experiment to find out how quickly the smell spreads from a particular type of perfume.

You will need to consider how you will decide when the scent has reached a particular part of the room. Would the same group of experimenters always get the same results? Would you get the same results in a different room? Or in the same room on a different day? Try to take these problems into account when you design your experiment.

Get your experiment checked, and then carry it out. Record your results in the way you think best. Discuss what your results suggest to you about the speed at which scent particles (molecules) move.

Questions

1   Using examples, and diagrams if they help, explain how each of the following supports the idea that everything is made up of particles:
   a Brownian motion
   b crystal structure
   c diffusion

Diffusion of two gases

This experiment will be demonstrated for you, as the liquids used should not be touched.

A piece of cotton wool is soaked in hydrochloric acid. A second piece of cotton wool is soaked in ammonia solution. The two pieces of cotton wool are pushed into the ends of a long glass tube. Rubber bungs are then pushed in, to seal the ends of the tube.

The hydrochloric acid gives off hydrogen chloride gas. The ammonia solution gives off ammonia gas. (Both of these gases smell very unpleasant, and should not be breathed in in large quantities.)

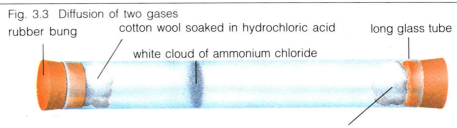

Fig. 3.3 Diffusion of two gases
rubber bung — cotton wool soaked in hydrochloric acid — long glass tube
white cloud of ammonium chloride
cotton wool soaked in ammonia solution

Questions

1 Hydrogen chloride and ammonia react together to form a white substance called ammonium chloride. Nearest which end of the tube does the ammonium chloride form?

2 How had the two gases travelled along the tube?

3 Which gas travelled faster?

4 The molecules of ammonia are smaller and lighter than the molecules of hydrogen chloride. What does this experiment suggest about how the size and mass of its molecules might affect the speed of diffusion of a gas?

5 If this experiment could be repeated at a higher temperature, would you expect it to take a longer or shorter time for the ammonium chloride to form? Explain your answer.

How small are potassium permanganate particles?

1 Measure $10 \, cm^3$ of water into a test tube. Add a few crystals of potassium permanganate and stir to dissolve.

2 Using a syringe, take exactly $1 \, cm^3$ of this solution and put it into a second tube. Add $9 \, cm^3$ of water to this tube, to make the total up to $10 \, cm^3$. You have diluted the original solution by 10 times. Put your two solutions side by side in a test tube rack.

3 Now dilute the second solution by 10 times, by taking exactly $1 \, cm^3$ of it, and adding it to $9 \, cm^3$ of water in a third tube. Add this tube to the row in the test tube rack.

4 Continue diluting the solution by 10 times, until you have a tube in which you can only just see the colour.

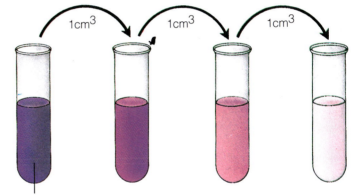

$1cm^3$ $1cm^3$ $1cm^3$

potassium permanganate solution made by dissolving a few crystals in $10cm^3$ of water

Fig. 3.4 Making serial dilutions of a potassium permanganate solution

Questions

Think about the number of potassium permanganate particles in each of your tubes. You began with a lot of particles in your first tube – all the ones that were in the crystals that you added to the water. You mixed them thoroughly into the water, and then took **one tenth** of them out to put into the second tube. So the second tube contained only one tenth as many potassium permanganate particles as the first one.

1 How many times fewer particles are there in the third tube than in the first tube?

2 How many times fewer particles are there in the last tube than in the first tube?

3 In your last, very faintly coloured tube the colour is still evenly spread through the water. So there must still be at least a few thousand potassium permanganate particles there. If you imagine that there are a thousand potassium permanganate particles in this tube, can you work out how many there must have been in the first tube? (You need to do some multiplications by 10 – lots of them.)

4 What does this experiment tell you about the size of potassium permanganate particles?

Questions

1 A petri dish was filled with agar jelly, in which some starch solution was dissolved. Two holes were cut in the agar jelly. Water was put into one hole. A solution of amylase was put into the other hole. Amylase is an enzyme which digests starch. It changes starch into sugar.

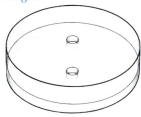

After a day, iodine solution was poured over the agar jelly in the dish. Iodine solution turns blue-black when in contact with starch. It does not change colour when in contact with sugar. The results are shown below.

hole containing water
hole containing amylase solution
petri dish

Results   starch agar

a Why did most of the agar jelly turn blue-black when iodine solution was poured over it?
b Why did the part of the jelly around the hole which contained amylase solution not turn black?
c The area which did not turn blue-black was roughly circular in shape. Explain why. Use the word 'diffusion' in your answer.
d Why do you think that water was put into one of the holes in this experiment?

2 Divide part of a page in your book into three columns, headed solid, liquid and gas. Copy each of the following words and statements into the correct column. Some of them may belong in more than one column.
a water at room temperature
b salt at room temperature
c oxygen at room temperature
d made up of particles
e particles vibrate
f particles move around freely, a long way apart
g particles vibrate slightly, and are held together tightly
h particles held together, but move around each other freely
i fills a container from the bottom
j completely fills a container
k does not spread out in a container
l particles move faster when heated
m particles move much closer together when squashed

3 Oxygen moves into the blood and carbon dioxide moves out of the blood in your lungs by diffusion.
a What is diffusion?
b Where is the concentration of oxygen highest – in your lungs, or blood arriving at your lungs?
c Using the words 'molecules' and 'concentration', explain why oxygen goes from your lungs into your blood.
d Breathing movements keep your lungs supplied with fresh air containing oxygen. This speeds up the rate at which the oxygen diffuses into your blood. Why?

4 Explain the following:
a If you grow a salt crystal it will be cube-shaped.
b A solution of potassium permanganate can be diluted many times, and still look purple.
c If ammonia solution is spilt in a corner of a laboratory, the people nearest it will smell it several seconds before the people in the opposite corner.
d Pollen grains floating on water appear to be jigging around if seen through a microscope.
e A copper sulphate crystal dissolves in water to form a blue solution. This happens faster in warm water than in cold water.

Gases get in and out of living things by diffusion

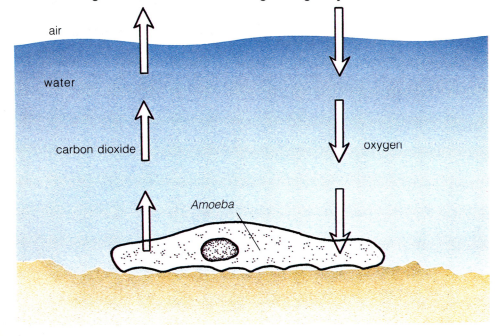

Diffusion is a very important process for living things. Oxygen diffuses from your lungs into your blood. Carbon dioxide diffuses from your blood into your lungs. Figure 3.5 shows how oxygen and carbon dioxide move in and out of a single-celled organism called *Amoeba*.

Fig. 3.5 Gas exchange in *Amoeba*. The *Amoeba's* cell uses oxygen in respiration. As the oxygen near to it is used up, this produces a low concentration of oxygen. The oxygen concentration in the air is greater so oxygen diffuses from the air into the *Amoeba*.
The *Amoeba* produces carbon dioxide in respiration. This produces a high concentration of carbon dioxide near the *Amoeba*. The concentration of carbon dioxide in the air is lower so carbon dioxide diffuses from the *Amoeba* into the air.

Question

The manufacture of a transistor

Diffusion is a process of great importance in the manufacture and operation of electronic components, such as transistors. A transistor is an electronic switch. It is used to control the flow of electrons in a circuit.

A transistor is formed from a silicon crystal, in which there is a small percentage of atoms of other substances. These other atoms alter the electrical properties of the crystal. The other atoms are added to particular regions of the silicon, so that these regions have different properties. These regions are known as n-type and p-type semiconductors.

Transistors need to be very small so that they can respond to changes quickly. There are several ways of making them.

This is one method.

First, a wafer of n-type silicon is heated in oxygen. The oxygen forms a layer of silicon oxide on the surface of the silicon. Chemicals are then used to etch away the oxide in a small area, of diameter a. In this area, the pure silicon is exposed. This is called an n-type semi-conductor.

Next, the silicon wafer is placed in a hot boron atmosphere. The boron atoms cannot penetrate the silicon oxide layer, but they can diffuse into the exposed silicon. They enter the surface and spread a little way under the silicon oxide layer. Wherever the boron enters the silicon, a p-type semiconductor is formed. The p-type semiconductor has a diameter a little larger than **a**.

Now the silicon wafer is heated in oxygen again to cover it completely with an oxide layer. A new area, of diameter b, is etched in the oxide. The wafer is then exposed to hot phosphorus gas. Phosphorus atoms diffuse into the exposed surface, but not through the silicon oxide. Where the phosphorus atoms enter the silicon, an n-type semiconductor is formed. The phosphorus atoms spread a little way under the oxide layer.

The junctions between the n-type and p-type semiconductors are very important if the transistor is to work well. By making them like this, the junctions are protected underneath a layer of silicon oxide.

Finally, electrical contacts are made by depositing metallic vapour on certain parts of the transistor and heating it. So now the whole transistor is covered and protected by either an oxide layer or the electrical contacts. This prevents impurity atoms in the air from diffusing into the junctions and spoiling the transistor.

Fig. 3.6

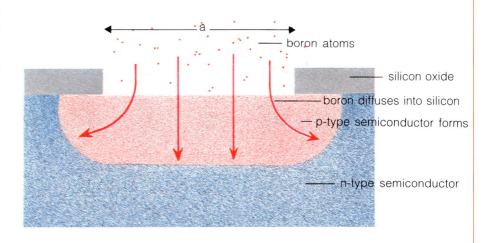

boron atoms
silicon oxide
boron diffuses into silicon
p-type semiconductor forms
n-type semiconductor

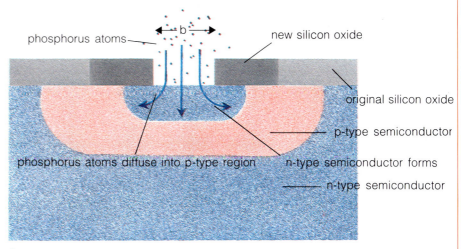

phosphorus atoms
new silicon oxide
original silicon oxide
p-type semiconductor
phosphorus atoms diffuse into p-type region
n-type semiconductor forms
n-type semiconductor

Questions

1 Why is it a good idea to make transistors small?

2 a How is a layer of silicon oxide made to form on the surface of the silicon wafer?

   b Explain how this layer helps in the next stage of the process.

3 Why are the new semiconductor regions bigger than the exposed areas on the silicon wafer?

4 The higher the temperature while the boron is entering the silicon wafer, the bigger the diameter of the p-type semiconductor region. Why?

5 What use is the oxide layer in the completed transistor?

4 MELTING AND BOILING

As a substance is heated it melts and then boils. The melting and boiling points of a pure substance are always the same, under the same conditions.

Heating substances makes the particles move faster

When a solid is heated, the particles (atoms or molecules) vibrate faster and faster. The energy of motion, or **kinetic energy**, of the particles increases.

Eventually, they have so much energy that they can begin to break away from each other. They begin to separate and move more freely. The solid **melts**, turning into a liquid.

If you go on heating the liquid, the heat energy makes the particles move even faster. As their kinetic energy increases, the clusters of particles begin to break apart. The individual particles move further away from each other. The liquid boils, turning into a gas. Boiling is when heat causes bubbles of vapour to form within a liquid.

Fig. 4.1 Atoms or molecules move faster when a substance is heated.

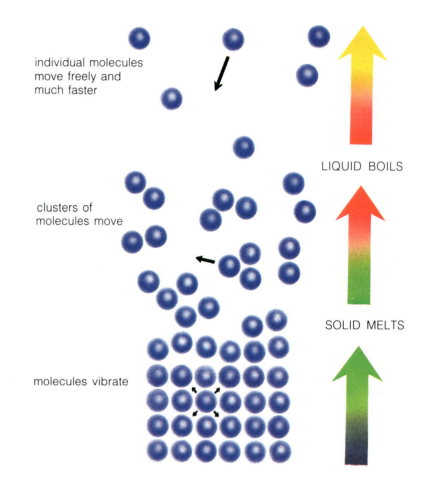

individual molecules move freely and much faster

LIQUID BOILS

clusters of molecules move

SOLID MELTS

molecules vibrate

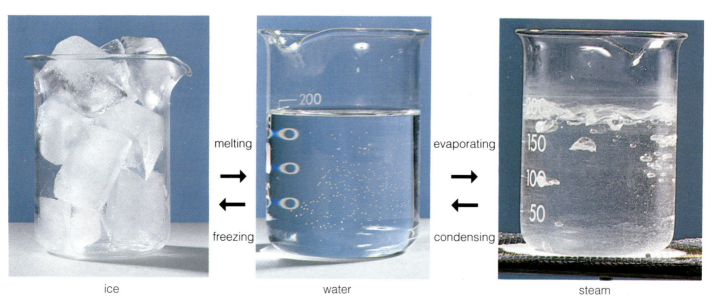

melting

freezing

evaporating

condensing

ice   water   steam

Fig. 4.2 Water changes from solid, through liquid, to gas when heated.

A pure substance has a constant melting and boiling point

When a solid melts, or a liquid boils, the bonds which hold the particles together are broken. The stronger the bonds are, the more energy is needed to break them. The stronger the bonds are, the higher the temperature must be to break them.

The strength of the bond between particles is not the same for different substances. The bonds will usually be broken at different temperatures. So different substances melt and boil at different temperatures.

A pure substance will always melt or boil at the same temperature under the same conditions. Pure water, for example, always melts at 0 °C and boils at 100 °C, at normal atmospheric pressure.

Some materials change directly from a solid to a gas

iodine vapour

sublimation

iodine solid

Fig. 4.3 Iodine changes directly from a solid to a gas when heated.

Some materials may change directly from a solid to a gas. This is called **sublimation**. Solid carbon dioxide changes to carbon dioxide gas as it warms up. It is called 'dry ice', and is often used to make 'mist' in theatre productions. Another substance which sublimes is iodine. Iodine crystals change to a purple gas when heated.

Melting and boiling points can be used for checking purity or temperature

If you have a sample of a substance and you want to find out if it is pure, you can measure its melting or boiling point. If it is not what it should be, then you know that your substance is not pure.

If you have a sample of a substance which you know is pure, you can use it to check a temperature. How could you do this?

Temperature does not rise while a substance is melting or boiling

When a solid is heated, its particles move faster and faster. The temperature of the solid steadily rises.

When the temperature reaches the melting point of the solid, the bonds between the particles start to break. While this is happening the temperature of the substance does not rise. You have to carry on heating until the bonds are all broken. Although you are putting in extra heat energy, the temperature does not rise. The extra energy is used to break the bonds.

When the bonds are broken, the solid melts and becomes a liquid. Now the temperature begins to rise again. The same thing happens while the liquid is boiling. When the temperature of the liquid has reached boiling point, it stays the same until the liquid has boiled. Once it is a gas the temperature can begin to rise again.

Question

A bowl of crushed ice was taken out of a freezer, and left on a laboratory bench to thaw. Its temperature was taken every 5 min for 1 h. The temperature of the laboratory was 20 °C.
The results are shown in the table below.

Time (min)	0	5	10	15	20	25	30	35	40	45	50	55	60
Temp. (°C)	-10	-5	-1	0	0	0	2	9	14	18	19	20	20

a Draw a line graph to show these results.
b What was the temperature inside the freezer?
c What was happening to the ice between 15 and 25 min after being taken out of the freezer?
d Why did the temperature not change between these times?
e Was the ice pure or did it contain impurities? Explain your answer.
f Would you expect the temperature of the water to continue rising above 20 °C? Explain your answer.
g A sample of benzene that had been cooled was put into a tube and heated in a water bath. Its temperature was taken every 5 min for 1 h. The results are shown below.

Time (min)	0	5	10	15	20	25	30	35	40	45	50	55	60
Temp. (°C)	5	5	5	13	28	45	60	73	78	80	80	80	80

h Plot these results as a line graph.
i What are the melting and boiling points of benzene?

When salt is put on roads, it stops ice from forming

Pure water freezes, and pure ice melts, at 0 °C. If there is an impurity in the water, it freezes at a lower temperature. An impurity lowers the freezing point of a substance.

If salt is spread on roads, the salt and water mixture will have a lower freezing point than pure water. Its freezing point is below 0 °C. So ice will not form until the temperature drops well below 0 °C. The more salt is added, the lower the freezing point becomes. You could put so much salt on the roads that no ice would form until the temperature became –18 °C! But this would not be a good idea, because salt on cars speeds up rusting. It also kills plants growing on the roadside verges. Salt is usually spread on roads when ice or snow is expected, to prevent the formation of a slippery surface. It is usually rock salt which has grit mixed with it. The grit helps to increase friction on the road. This gives tyres a better grip, even if some ice does form.

INVESTIGATION 4.1

Cooling wax

This experiment tests your powers of observation. You have probably seen melted wax cooling before. This time, really watch it carefully!

1 Put some wax into a test tube, and stand the tube in a beaker of boiling water. Leave it until the wax has melted.

2 Now let the wax cool down slowly. Watch it carefully. Record anything that appears to be interesting.

3 Try to explain everything that you see. You will need to think about particles (molecules) of wax.

Fig. 4.4   A gritting lorry. Salt and grit spread on roads prevent ice forming, and improve grip.

INVESTIGATION 4.2

Measuring the melting point of a solid

1 Set up the apparatus shown in Figure 4.5.

2 Draw a results chart, so that you can fill in your readings from instruction 4.

3 Heat the water until the solid melts.

4 *Work quickly.* Take the tube, containing the melted substance, out of the beaker of hot water. Put it into a clamp on a retort stand. Record its temperature every 30 s. Do this until the substance has completely solidified.

5 Plot a cooling curve for the substance. Put time on the horizontal axis, and temperature on the vertical axis.

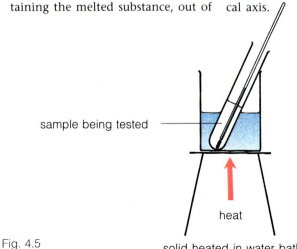

sample being tested

heat

solid heated in water bath

Fig. 4.5

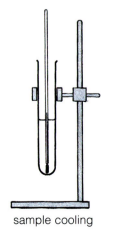

sample cooling

Questions

1   Why do you think that a water bath was used to heat the substance?

2   This substance has a melting point below 100 °C. Would this method work for a solid with a melting point higher than this? How could you adapt the apparatus to make it suitable for a solid with a higher melting point?

3   Why was the tube supported in a clamp as it cooled, and not left in the beaker of water?

4   What is the melting point of this substance?

5   Find out what the melting point of a pure sample of this substance should be. Was your sample pure?

6   Why is this experiment not accurate? Suggest some ways in which it could be made more accurate.

Questions

1 The graph shows the melting and boiling points of five substances.
  a Which substance has the lowest boiling point?
  b What is the melting point of:
  i ethanol ii mercury iii oxygen?

Room temperature is around 20 °C.
  c Which of these substances is/are solid at room temperature?
  d Which is/are liquid at room temperature?
  e Which is/are gaseous at room temperature?

2 A sample of water is tested to see if it is pure. It is found to boil at 104 °C.
  a Is the water pure?
  b Will it freeze at 0 °C, above this, or below this temperature? Explain.

3 Cockroaches can be frozen in ice and revived without any apparent harm. They have a type of antifreeze in their blood and cells.
  a How might ice crystals damage a cockroach's cells?
  b Why might it be an advantage for a cockroach to have antifreeze in its body?

4 a If a pond is covered with a layer of ice, where in the pond would the warmest water be found?
  b What would the temperature of this water be?
  c If there were fish in the pond, why would it be important to keep a hole in the ice?
  d A concrete-lined pond may crack in a cold winter. Why?

5 Ethylene glycol is used as an antifreeze in car engines. When it is mixed with water, it lowers the freezing point of the water. The mixture is used in the car engine's cooling system. The chart shows the freezing points of mixtures of ethylene glycol and water.
  a Plot this information as a line graph:

% water	100	90	80	70	60	50	40
% ethylene glycol	0	10	20	30	40	50	60
freezing point (°C)	0	–4	–9	–16	–24	–34	–47

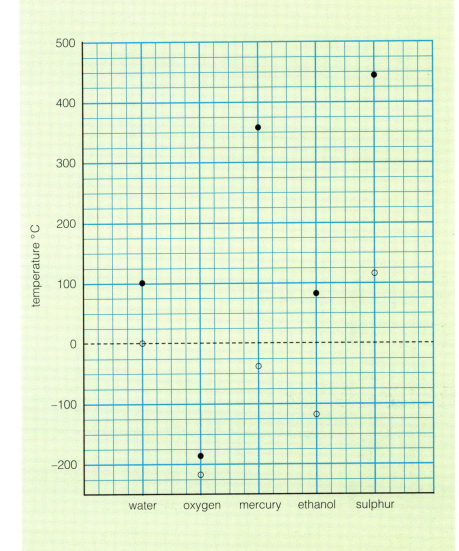

● Boiling Point
○ Melting Point

temperature °C

water   oxygen   mercury   ethanol   sulphur

b When water changes to ice, it expands. How could this damage a car's cooling system?
c What is the freezing point of pure water?
d What is the freezing point of a 45: 55 water:ethylene glycol mixture?
e What mixture would you use in your car radiator if the lowest winter temperature was expected to be –26 °C?
f Your car's cooling system has a capacity of 7000 cm³. Use your answer to question e to calculate how much ethylene glycol you need to add to the cooling water.

19

5 ATOMIC STRUCTURE

Atoms themselves are made up of even smaller particles. The most important of these are protons, neutrons and electrons.

Atoms contain protons, neutrons and electrons

Fig. 5.1   An atom

nucleus, containing neutrons and protons

region where electrons are found

Atoms are made up of even smaller particles. There is a **nucleus** in the centre of each atom. The nucleus can contain **protons** and **neutrons**. Around the nucleus is the rest of the atom, where **electrons** are most likely to be found. Protons and neutrons have about the same mass. Electrons are about 2000 times lighter.

Protons have a small positive electrical charge. Electrons have an equal but opposite (negative) charge. The number of protons and electrons in an atom are exactly equal, so the two equal and opposite charges cancel out. Atoms have no overall charge.

Questions

1   If an atom has 10 protons, how many electrons does it have?
2   Which particle in an atom carries a positive charge?
3   Which particle in an atom carries a negative charge?

Protons, neutrons and electrons

Particle	Relative mass	Relative charge
proton	1	+1
neutron	1	0
electron	1/2000	-1

The history of atomic structure

The Ancient Greeks were the first people to think of matter as being made of tiny particles. They imagined these particles to be like solid balls and this idea of the atom remained until the early 1900s. Around this time, evidence emerged that atoms contained at least two kinds of matter, some of it with a positive charge and some with a negative charge. A new model of the atom was suggested – a ball of positively charged 'dough' in which negatively charged electrons were dotted around like currants.

Fig. 5.2a The ancient Greeks imagined that atoms were like solid balls.

Fig. 5.2b Around 90 years ago it was suggested that an atom was rather like a plum pudding.

The Rutherford experiment

Ernest Rutherford was born in New Zealand in 1871. He won a Nobel Prize in 1908 for work on radioactivity. While continuing this work in Manchester, England, he made a discovery which he, and all other scientists at the time, found quite amazing.

He and two colleagues, Geiger and Marsden, were carrying out an experiment in which they shot alpha particles at a very thin piece of gold foil, in a vacuum. The apparatus is shown in Figure 5.3a.

Most of the alpha particles went straight through (A), some went through but changed direction slightly (B) and an even smaller number actually bounced back (C).

This suggested to Rutherford that the atom must be mainly space, and that the positive charge was not spread around, but in the centre. There was a positively charged central core, made of particles called protons, with the negatively charged electrons around the outside. He put forward this new model of the atom in 1911, and the modern view of the atom is very similar.

Fig. 5.3b Particles can pass through the gold because most of an atom is space.

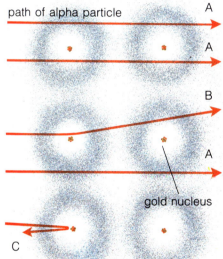

path of alpha particle

A
A
B
A

gold nucleus

C

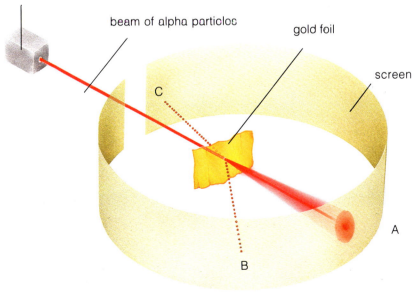

alpha particle source

beam of alpha particles

gold foil

screen

C

A

B

Fig. 5.3a Rutherford's experiment. The entire apparatus was enclosed in a vacuum chamber. Why do you think this was necessary?

Rutherford was astonished at these results. He wrote: 'It was quite the most incredible event that ever happened to me in my life. It was as incredible as if you fired a 15-inch shell at a piece of tissue paper and it came back and hit you. On consideration, I realised that this scattering backwards must be the results of a single collision, and when I made calculations I saw that it was impossible to get anything of that order of magnitude unless you took a system in which the greater part of the mass of the atom was concentrated in a minute nucleus.'

6 ELEMENTS

An element is a substance made of atoms which all contain the same number of protons.

There are over 100 different kinds of atom

Over 100 different kinds of atoms exist. They are different from each other because they do not have the same numbers of protons, neutrons and electrons.

Atoms with the same number of protons as each other behave in the same way chemically. Atoms with different numbers of protons behave differently. There are about 90 naturally occurring types of atoms. The remainder are made by humans.

Fig. 6.1 Some examples of elements. Beginning at the back right, and working clockwise: chlorine (in the gas jar), chromium, sulphur, iodine, zinc, carbon, and mercury.

An element is a substance whose atoms all have the same number of protons

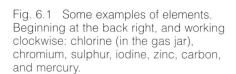
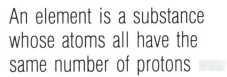

A substance made from atoms which all have the same number of protons is called an **element**. As there are about 90 naturally occurring kinds of atoms, there are about 90 naturally occurring elements. You can see them all listed in the Periodic Table, on page 27. Each element has its own symbol. Hydrogen, for example, has the symbol **H**. Helium has the symbol **He**.

The atomic number of an element is the number of protons in each atom

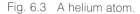

The number of protons an atom contains is called its **atomic number**. All the atoms of an element contain the same number of protons so they all have the same atomic number. The element with the smallest number of protons is hydrogen. It has just one proton, so its atomic number is one. Helium has two protons, so its atomic number is two.

You sometimes need to write the atomic number of an element when you write its symbol. You show it like this: $_1$**H**. This shows that the atomic number of hydrogen is one. Helium has the atomic number two so it is written $_2$**He**.

Fig. 6.2 A hydrogen atom.

Atomic number = number of protons = 1

Mass number = number of protons + number of neutrons = 1

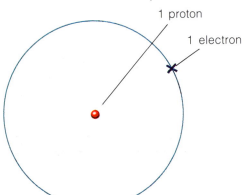

Fig. 6.3 A helium atom.

Atomic number = number of protons = 2

Mass number = number of protons + number of neutrons = 4

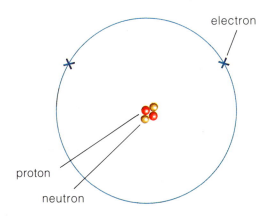

The mass number of an atom is the total number of protons and neutrons it contains

The **mass number** of an atom is the total number of 'heavy' particles that it contains. If you look at the table on page 20, you will see that the 'heavy' particles are neutrons and protons. A hydrogen atom contains just one proton, one electron and no neutrons, so its mass number is one. It is the smallest atom.

The next largest atom is the **helium** atom. A helium atom contains two protons, two neutrons and two electrons. As it has two protons, its atomic number is two. It has four 'heavy' particles – two protons and two neutrons – so its mass number is four.

If you want to show the mass number of an element, you write it like this: **¹H** or **⁴He**.

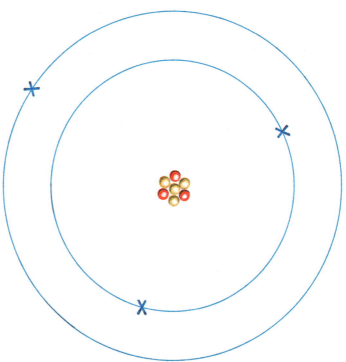

Fig. 6.4 A lithium atom. Lithium is the next largest atom after helium. What is its atomic number? What is its mass number?

Fig. 6.5 Using symbols to show atomic number and mass number.

$$^{235}_{92}\text{U}$$

This is the mass number. It tells you the total number of protons and neutrons in one atom of the element.

This is the atomic number. It tells you how many protons there are in one atom of the element.

Questions

1. What is an element? Give three examples of elements.
2. Use the table on page 25 to write down the symbols of these elements:
   a carbon b chlorine c oxygen
   d sodium e potassium f helium
   g nitrogen h neon i argon j boron
   k sulphur l magnesium
3. a What is meant by the term atomic number?

   b What is the atomic number of:
   i hydrogen
   ii lithium
   iii uranium?

   c What is meant by the term mass number?

   d For each of the following examples, give (a) the atomic number and (b) the mass number:
   i $^{35}_{17}\text{Cl}$ **ii** $^{12}_{6}\text{C}$ **iii** $^{14}_{6}\text{C}$ **iv** $^{16}_{8}\text{O}$

4. Copy and complete the following table.

Mass number	20	40			28
Atomic number	10			13	
Number of neutrons		22			
Symbol	Ne	Ar	$^{24}_{12}\text{Mg}$	$^{27}_{13}\text{Al}$	$^{28}_{14}\text{Si}$

7 ELECTRON ORBITS

The electrons of an atom are arranged in orbits around the nucleus. Each orbit can only hold a certain number of electrons.

Electrons are arranged in orbits

You have seen how different kinds of atoms have different numbers of protons. A hydrogen atom has one proton. A lithium atom has three protons. A sodium atom has 11 protons. The number of protons in an atom is called its atomic number. The number of protons in an atom determines what sort of atom it is.

The number of electrons in an atom is the same as the number of protons. So a hydrogen atom has one electron. A lithium atom has three electrons. A sodium atom has 11 electrons. The electrons are arranged in **orbits** around the nucleus of the atom. The orbits are sometimes called shells.

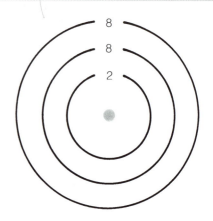

Fig. 7.1 Electron orbits. The inner orbit can hold up to two electrons. The next two can each hold eight.

Each orbit can only hold a certain number of electrons

The electron orbits are filled up from the inside outwards. The first orbit can hold up to two electrons. The second orbit can hold up to eight electrons. The third orbit can also hold up to eight electrons.

A lithium atom has three protons and three electrons. Two electrons are in the first orbit, so this orbit is full. The third electron is in the second orbit. A sodium atom has 11 protons and 11 electrons. Two electrons fill up the first orbit. Eight electrons fill up the second orbit. The last electron goes into the third orbit.

The arrangement of electrons is called the electron configuration

The way in which the electrons are arranged in an atom is called its **electron configuration**. You can show the electron configuration by drawing the atom. Or you can show it by numbers. For example, the electron configuration of a lithium atom is **Li (2,1)**. This means that lithium has two electrons in its inner orbit, and one electron in the second orbit. The electron configuration of a sodium atom is **Na (2,8,1)**.

The electron configuration of an atom determines how that atom will react with other atoms. You will find much more about this later in the book.

Fig. 7.2 The electron configurations of three atoms.

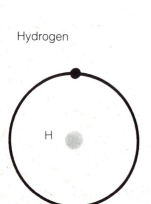

Hydrogen

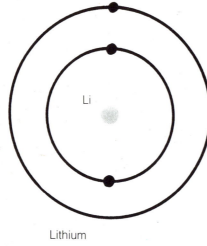

Lithium

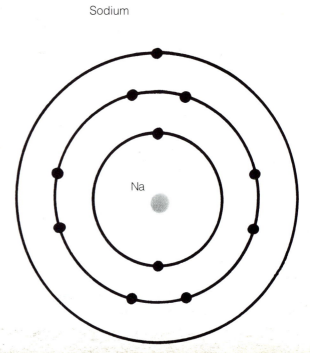

Sodium

Element	Symbol	Atomic number (number of protons)	Mass number (number of protons + neutrons)	Electron configuration
hydrogen	H	1	1	1
helium	He	2	4	2
lithium	Li	3	7	2,1
beryllium	Be	4	9	2,2
boron	B	5	11	2,3
carbon	C	6	12	2,4
nitrogen	N	7	14	2,5
oxygen	O	8	16	2,6
fluorine	F	9	19	2,7
neon	Ne	10	20	2,8
sodium	Na	11	23	2,8,1
magnesium	Mg	12	24	2,8,2
aluminium	Al	13	27	2,8,3
silicon	Si	14	28	2,8,4
phosphorus	P	15	31	2,8,5
sulphur	S	16	32	2,8,6
chlorine	Cl	17	35	2,8,7
argon	Ar	18	40	2,8,8
potassium	K	19	39	2,8,8,1
calcium	Ca	20	40	2,8,8,2

Table 7.1   The first twenty elements.

Questions

1 Draw the following atoms to show the number of electrons in each of their electron orbits:
a carbon b oxygen c neon
d chlorine e argon f potassium

2 Name three elements whose atoms have full outer electron orbits.

3 Name three elements whose atoms have only one electron in their outer orbit.

4 Name three elements whose atoms have outer orbits which are one electron short of being full.

8 THE PERIODIC TABLE

The different kinds of elements can be arranged in a pattern, according to the structure of their atoms and the way in which they behave.

Elements can be arranged into groups

Each element behaves in a different way from every other element. We say that their **properties** are different. But there are also similarities in the way that some elements behave.

For example, you will see in Topic 80 that **lithium**, **sodium**, and **potassium** have very similar properties. They are all reactive metals. They are often called the **alkali metals**.

Fluorine, **chlorine**, **bromine** and **iodine** also show great similarities in the way they behave. They are often known as the **halogens**.

Elements with a similar arrangement of electrons in their outer orbit behave in a similar way

When chemists first arranged elements into groups, they did not know why some elements showed similarities in the way they behaved. Now we know that it is to do with the number of electrons they have in their outer orbit. Elements with the same number of electrons in their outer orbit behave in a similar way.

For example, lithium, sodium and potassium all have **one** electron in their outer orbit. Fluorine, chlorine, bromine and iodine all have **seven** electrons in their outer orbit. Lithium, sodium and potassium belong to **Group 1**, because their outer orbit contains one electron. Fluorine, chlorine, bromine and iodine belong to **Group 7**, because their outer orbit contains seven electrons.

The Periodic Table shows all the elements arranged in Groups

You can see these Groups in the Periodic Table. The elements in each Group are arranged vertically. The element with the smallest atomic number is at the top of the Group, and the one with the largest atomic number is at the bottom.

The horizontal rows in the Periodic Table are called **Periods**. Look at Period 2. It begins with lithium, then beryllium, then boron. These elements are arranged in increasing atomic number. (Remember – the atomic number is the number of protons in an atom.)

The Periodic Table can suggest how reactive an element will be

Atoms are most stable when their outer electron orbit is full. The Group 0 elements have eight electrons in their outer orbit. The orbit is full. So Group 0 elements are very **unreactive**. They are sometimes called the **noble gases**.

Group 7 elements have seven electrons in their outer orbit. They only need one more electron to fill this orbit. So they readily take electrons from other atoms, to fill up their outer orbit and become stable. This makes them very **reactive** elements. You will find more about this in Topic 81.

Group 6 elements have six electrons in their outer orbit. They need two more electrons to fill this orbit. Like Group 7 elements, they will take electrons from other atoms. But they do this less readily, so they are less reactive than Group 7 elements.

At the other end of the Periodic Table, Group 1 elements have only one electron in their outer orbit. They can have a full outer orbit by losing this electron. They do this very easily, giving up their electron to other atoms. So Group 1 elements are very reactive.

Apart from the noble gases, the most reactive elements are near the left- and right-hand sides of the Periodic Table.

Metals are on the left-hand side of the Periodic Table

The zig-zag line separates metals from non-metals. All the elements on the left-hand side of the line are metals. All the elements on the right-hand side of the line are non-metals.

Metals are elements which tend to lose electrons. The metals have atoms which can most easily end up with full electron orbits by losing electrons.

The most reactive metals are the ones nearest the left-hand side of the Periodic Table, such as sodium and potassium. Why do you think that these are the most reactive metals?

Transition elements have more complex electron arrangements

The middle of the Periodic Table is taken up with the **transition elements**. You will find out more about some of these elements on page 40. These elements do not fit into one of the eight groups. They have more complex electron arrangements.

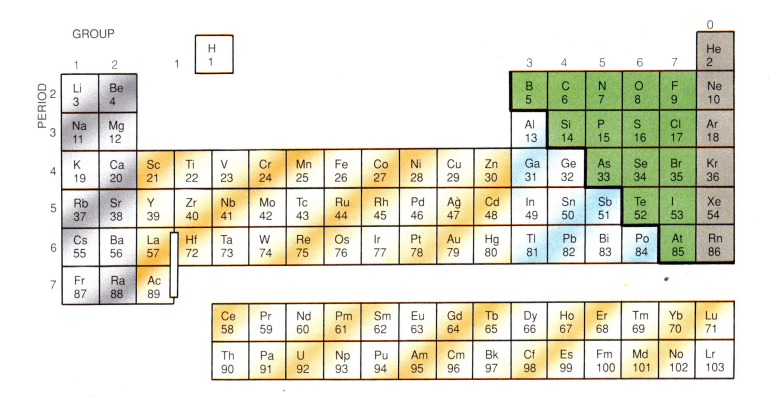

KEY

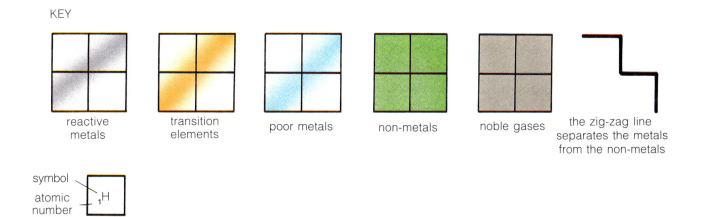

reactive metals

transition elements

poor metals

non-metals

noble gases

the zig-zag line separates the metals from the non-metals

symbol
atomic number

₁H

hydrogen

Fig. 8.1 The Periodic Table. The Periodic Table lists all the elements in order of their atomic number. The elements are arranged in Groups, according to the number of electrons in their outer orbit. For example, Group 1 elements – Li, Na, K, Rb, Cs and Fr – all have one electron in their outer orbit. All the elements beyond uranium (atomic number 92) have been made by people. They do not occur naturally on Earth.

Questions

1 What is the atomic number of :
   a sodium
   b oxygen
   c copper?

2 Which element has the atomic number of:
   a 17
   b 6
   c 11?

3 How many electrons are there in the outer orbit of:
   a a magnesium atom
   b a neon atom
   c a nitrogen atom?

4 a Why are the elements in Group 1 very reactive elements?
   b Why are the elements in Group 7 very reactive elements?
   c Why are the elements in Group 0 very unreactive elements?

— EXTENSION —

Isotopes of an element have the same atomic number, but different mass numbers

All atoms with the same number of protons belong to the same element. They behave in exactly the same way in chemical reactions. As well as having the same number of protons, they also have the same number of electrons. For example, all hydrogen atoms have one proton and one electron. The atomic number is one.

Most hydrogen atoms have no neutrons so the mass number is one. But about one hydrogen atom in 10 000 is heavier than this. It has a neutron in its nucleus. This form of hydrogen still has one proton. It still has an atomic number of one. It is chemically identical to the most common form. It occupies the same place in the Periodic Table. For this reason it is called an **isotope** of hydrogen. ('Isotope' means

'same place'.) This isotope is sometimes called **deuterium**. Hydrogen and deuterium are both isotopes of hydrogen. They have the same atomic number, but different mass numbers. A

deuterium atom is heavier than a hydrogen atom, because of its extra neutron.

Fig. 9.1a   A hydrogen atom

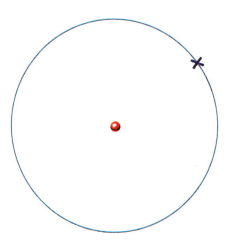

Fig. 9.1b   A deuterium atom

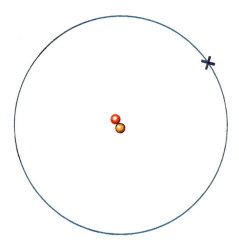

Carbon also has isotopes

Hydrogen is not the only element to have isotopes. Many elements have isotopes. Carbon has several isotopes. The 'normal' carbon atom has six protons and six neutrons. It has a mass

number of 12, and its symbol is ^{12}C. It is called carbon twelve. Another isotope has eight neutrons. It still has six protons, or it would not be carbon. Its mass number is 14 and its symbol is ^{14}C. It is called carbon fourteen.

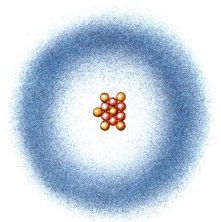

Fig. 9.2a   A carbon 12 atom has 6 protons and 6 neutrons.

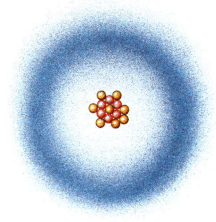

Fig. 9.2b   A carbon 14 atom has 6 protons and 8 neutrons.

Relative atomic masses are used for the masses of atoms

A single atom of hydrogen has a mass of 0.00000000000000000000000017g. This is a rather clumsy number to use. Instead of using grams, we compare the masses of atoms to the mass of a single atom of carbon 12. Carbon 12 has a mass of 12 atomic mass units. A hydrogen atom has a mass one twelfth of this. The atomic mass of hydrogen is one. The atomic mass of carbon 14 is 14 atomic mass units. Chromium has a mass which is twice that of carbon 12. Its atomic mass is 24.

Mass number is the atomic mass of a particular isotope of an element

When we talk about the atomic mass of an element, we can mean two things. We might mean the atomic mass of a particular isotope of that element. Or we might mean the average atomic mass of all the isotopes present in a particular sample of that element.

Take the element carbon, for example. The commonest isotope of carbon has six protons and six neutrons. We can say that its mass number is 12. Mass number refers to the atomic mass of a particular isotope of an element. If you say 'the mass number of carbon is 12', you really mean 'the mass number of the commonest isotope of carbon is 12'.

Relative atomic mass is an average for the different isotopes

In a sample of carbon, there will be atoms with different mass numbers. If you took 100 carbon atoms, you would probably find that 99 of them were carbon 12 atoms. One of them would probably be a carbon 13 atom. If you sorted through millions of carbon atoms, you might find a carbon 14 atom.

So if you found the *average* mass of a random sample of 100 carbon atoms, it would be a little bigger than 12. The few heavier atoms would make the average mass about 12.01. This average mass number of a random sample of an element, taking into account the different isotopes in it, is called the **relative atomic mass**. Because the relative atomic mass of an element is an average, it is often not a whole number.

Chlorine, for example, has two common isotopes. They are chlorine 35 and chlorine 37. In a sample of chlorine, there are nearly three times as many chlorine 35 atoms as chlorine 37 atoms. This makes the average mass of the atoms 35.5. The relative atomic mass, or A_r, of chlorine is 35.5.

Table 9.1 Some relative atomic masses

element	relative atomic mass A_r	most common isotope	percentage occurrence
chlorine	35.5	^{35}Cl	75
barium	137.34	^{138}Ba	72
germanium	72.69	^{74}Ge	36.5
mercury	200.5	^{202}Hg	30
strontium	87.62	^{88}Sr	82.6
thallium	204.3	^{205}Tl	70.5

Questions

1 a What do all isotopes of a particular element have in common?

 b How do isotopes of a particular element differ from each other?

2 Three isotopes of magnesium are ^{24}Mg, ^{25}Mg and ^{26}Mg.

 a Use the Periodic Table on page 27 to find the atomic number of magnesium.

 b Every magnesium atom has the same number of protons. What is this number?

 c What is the total number of protons and neutrons in a ^{24}Mg atom?

 d How many neutrons are there in a ^{26}Mg atom?

 e The relative atomic mass of magnesium is 24.3. Why is this not a whole number?

 f Which is the most common isotope of magnesium?

3 In 1000 thallium atoms, there are 705 atoms of ^{205}Tl.

 a What is the total number of neutrons and protons in these 705 thallium atoms?

 b The remaining 295 atoms are ^{203}Tl atoms. What is the total number of neutrons and protons in these 295 thallium atoms?

 c What is the average number of protons and neutrons in the 1000 thallium atoms?

 d What is the A_r (relative atomic mass) of thallium?

4 Natural copper contains two isotopes, ^{63}Cu and ^{65}Cu. The atomic number of copper is 29.

 a How many protons are there in a copper atom?

 b How many neutrons are there in each of the two isotopes of copper?

 c In naturally occurring copper, 69% of the atoms are ^{63}Cu, and the remainder are ^{65}Cu. What is the relative atomic mass of copper?

10 METALS AND NON-METALS

All elements can be put into two groups – metals or non-metals. Four-fifths of all elements are metals.

Most elements are metals

You will remember that there are about 90 naturally occurring elements. The Periodic Table on page 27 shows these elements arranged in Groups. All the elements to the left of the zig-zag line are **metals**, and all those to the right are **non-metals**. There are far more metallic elements than non-metallic ones. Around four-fifths of all the different elements are metals.

Fig. 10.1 Around four-fifths of all the known elements are metals. But metals only make up less than one quarter of the materials found on Earth, because many metals are very rare.

Metals and non-metals have different properties

Metals have certain properties in common. They are usually **shiny**. However, some metals, like iron, look dull because they have a covering of metal oxide on their surface. If you rub this off with sandpaper you can see the shiny metal beneath.

All metals will **conduct electricity**. Most non-metals will not. One exception is carbon, which is a non-metal but *can* conduct electricity.

Most metals **melt** when heated. Some of the more reactive ones, such as magnesium, also burn with a flame when you heat them. It is hard to generalise about non-metals, but some, like sulphur, phosphorus and carbon, **burn** when heated in air.

When substances burn in air, they form compounds called oxides. When a metal burns it forms an oxide that is **basic**. When a non-metal burns it forms an oxide that is **acidic**.

Most metals are **malleable**. This means that they can be beaten or rolled into different shapes. Non-metals tend to be brittle, which means that they break easily.

Fig. 10.2 Chromium and sulphur. Chromium is a metal and sulphur is a non-metal.

Question

You have probably seen examples of metals and non-metals in previous lessons. If so, try this question from memory. Copy this chart, and fill in the spaces. In each case, there are two statements to choose from.

Substance	Metal or non-metal?	Dull or shiny?	Good or poor conductor of electricity and heat?	Melts or burns when heated?	Brittle or malleable?	Is the oxide acidic or basic?
Silver						
Carbon						
Magnesium						
Copper						
Sulphur						

Metals have their special properties because of the atoms of which they are made

Metallic elements have atoms which lose some of their electrons quite easily. The reason they tend to do this is explained on page 36. In a piece of metal, the atoms are packed together in a regular pattern. They are packed so tightly that some of the electrons become detached from their atoms. These 'free' electrons move around in the spaces between the atoms. They are shared between the atoms in the piece of metal.

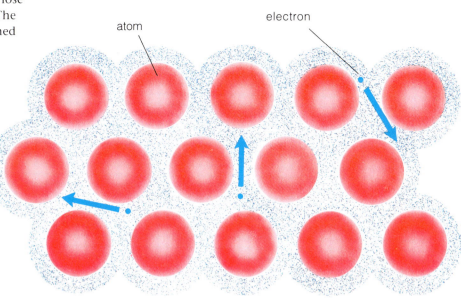

Fig. 10.3 How metals conduct heat and electricity. The atoms in a metal are packed in a regular pattern called a **lattice**. Some of the electrons can move freely in the spaces between the atoms. The freely-moving electrons can transfer heat and electricity through a piece of metal.

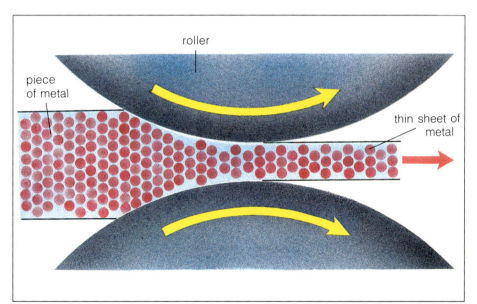

Fig. 10.4 Rolling a sheet of metal. Metals such as aluminium can be rolled into thin sheets, because the regular layers of atoms can slide over one another.

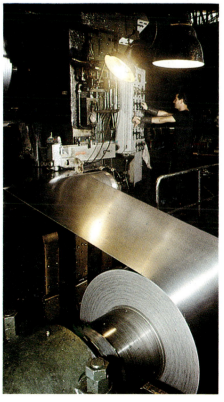

Fig. 10.5 Rolling sheet metal

The electrons carry a negative charge. The metal atoms carry a positive charge, because they now have more protons than electrons. The positive and negative charges attract each other, so the whole arrangement is held closely together.

The free electrons explain why metals are so good at conducting electricity. The electrons in metals can readily be made to move in a particular direction. This movement is what we call an electric current.

Heat is conducted through a material by the movement of its particles. The free electrons in a metal can easily vibrate when they get hotter, and bump into neighbouring electrons, making them vibrate as well. In this way heat is quickly passed from one end of a strip of metal to the other.

When a piece of metal is hammered it can easily change shape as the layers of atoms slide over each other. This is why metals are malleable. For the same reason, many metals can be drawn out into thin wires. They are said to be **ductile**.

11 USES OF METALS

The properties of metals make them very useful to us for making all sorts of things. Mixtures of two or more metals, called alloys, are also widely used in constructing all kinds of objects.

Each metal has its own properties

Although all metals share certain properties, no two metals are identical. These individual properties make different metals suitable for different uses. Sometimes there is a need for a substance which has a combination of properties not found in any one metal. By mixing two or more metals together a new substance called an **alloy** can be made. Sometimes an alloy consists of one or more metals mixed with a tiny amount of non-metal, e.g. steel. This new substance will have a different set of physical properties from either of its 'parent' metals.

Only some of the most commonly used metals and alloys are mentioned on these two pages. Engineers are constantly developing new alloys for new purposes.

Fig. 11.1 A technician in a metallurgy laboratory, testing the strength of metal structures.

Uses of some pure metals

One metal which you have almost certainly made use of is **aluminium**. Aluminium is a light metal which can easily be rolled out into very thin, yet strong sheets. It is also very resistant to corrosion. This makes it suitable for milk bottle tops. It does not melt easily and can withstand high temperatures, so it is used for making containers for prepacked foods, and for making foil for covering food as it cooks. Many drink cans are made of aluminium as it is non-corrosive, light and yet strong.

Copper is especially good at conducting electricity and is ductile. This means that it can be pulled out into a thin wire. So copper is used for making electrical wires.

Gold, **silver** and **platinum** are used for making jewellery. They are all very unreactive metals, so they do not tarnish easily and stay shiny and attractive. They may also be used for coating parts in electrical switches.

Mercury is an unusual metal because it is liquid at room temperature. It is used in thermometers.

Lead is a heavy metal which is extremely malleable. This makes it suitable for making waterproof edgings to roofs such as around chimneys. Before it was known to be poisonous, lead was used for making water pipes and toy soldiers. Lead is also used for shielding from radiation.

Fig. 11.2 Objects made from pure metals. They include lead, aluminium, iron, tungsten, silver, gold and copper. Can you identify which is which? What properties of each metal make them suitable for these purposes?

Uses of some alloys

The commonest alloy in use today is **steel**. Steel is an alloy of **iron** and **carbon**. The carbon makes the iron much stronger. There are different kinds of steel. They can have varying amounts of carbon in them, and many kinds have other elements as well. Steel which contains only iron and carbon is called **mild steel**. This is the sort of steel which is normally used for building large structures such as bridges and cars. For making cutlery, **nickel** and **chromium** are added as well. This makes the steel even harder and stops it rusting. It is called **stainless steel**.

The first alloy discovered by humans was **bronze**. Bronze is an alloy of **copper** and **tin**, and was used in the Bronze Age for making weapons. Later, when iron was discovered, bronze weapons rapidly became outdated because they were not very strong and did not stay sharp for long. But we still use bronze for statues because it is a very attractive alloy, and does not rust.

Solder is an alloy of two metals with low melting points – **tin** and **lead**. The alloy has an even lower melting point than either of its two 'parent' metals, and is used for joining metals together, such as wires.

Fig. 11.3   Objects made from alloys. They include brass, solder, stainless steel, aluminium alloy, and high-tensile steel. Can you identify which is which? Which metals are present in each alloy? What properties of each alloy make it suitable for the purpose shown?

Comparing the strength of different metals

You are going to test wires made from different metals, to find out how strong they are. You may also be able to test different thicknesses of wire of the same metal.

1 Read through the experiment and draw up a suitable results chart. You will need to record the type of metal used, its thickness, and the weight that it supported.

2 Put one piece of wire aside – you are not going to test this until the end of your experiment. It is best to choose a medium thick wire of a metal of which you have several other thicknesses.

3 Choose your first wire. Record the metal it is made from, and its thickness, in your table. Make it into a hook, by bending a short length of the wire round a pencil.

4 Mount your hook as shown in the diagram. Now hang masses on it until it gives way. (It is important to decide when you think the hook has given way, and to do the same each time.) Record the maximum weight it supported.

5 Now do the same with other metals, or with other thicknesses of the same wire. It is important that you make your hook in exactly the same way each time.

Questions

1 For each metal that you tested, plot a line graph of weight supported (vertical axis) against thickness of wire (horizontal axis).
2 Why was it important to bend each of your hooks by the same amount when you made them?
3 Which was the strongest hook?
4 Which was the strongest metal?
5 Use your graphs to predict the weight that your untested piece of wire could support. Write this down and then test it. How close was your prediction?
6 How accurate do you think your results are? Can you suggest any ways in which this experiment might be improved?

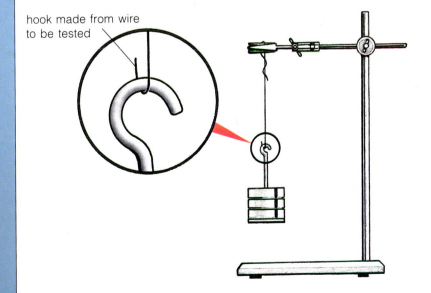

hook made from wire to be tested

Fig. 11.4

Breaking a wire

This experiment will probably be demonstrated, because when a wire suddenly breaks the ends can fly outwards with considerable force!

Watch carefully as a wire is loaded with weights. The weights will be added until the wire snaps. This experiment can be repeated with different types of wire.

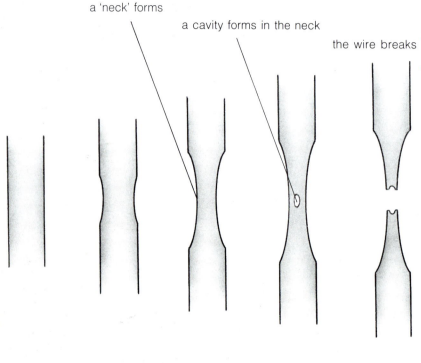

a 'neck' forms

a cavity forms in the neck

the wire breaks

Fig. 11.5 Breaking a wire

Questions

1   Does the wire snap suddenly, or can you predict when it will snap? If so, how can you predict it?
2   Are the two halves of the broken wire, added together, longer than the original?
3   Does the type of metal used make much difference to the result?
4   Is the strongest wire the same as that found in the hook experiment (Investigation 11.1)?
5   Arrange the wires in order of strength. Do the same for your results from the hook experiment. Are the orders the same?
6   Were your doing the same thing to the wires in this experiment, and in the hook experiment? Try to think what is happening to the metal atoms, and describe it.
7   Look very carefully at the broken ends of the wire. Use a hand lens. You may be able to see a small hole in each broken end. The diagram shows how this forms.
    What must be happening to the layer of atoms in the wire as this happens?

Questions

1   Which of these elements are metals, and which non-metals?
    a iron **b** nickel
    c carbon **d** oxygen
    e gold **f** helium
    g magnesium **h** sulphur
    i mercury **j** tin
2 **a** Make a list of four properties of metals.
  b Name two metals which are lacking one or more of these properties, and explain why they are unusual.
3   Why is copper a good conductor of electricity?

4 **a** What is an alloy?
  b Name two alloys. For each list its main uses, and the properties which make it suitable for these uses.
5   Lead and tin can be mixed together to form solder. The chart shows the melting points of alloys of different proportions of tin and lead.
  a Plot a line graph to show this data. Put percentage of tin on the horizontal axis, and melting point on the vertical axis.

  b Which has the lowest melting point – tin or lead?
  c What percentages of lead and tin would you mix to get an alloy with the lowest possible melting point? What would this melting point be?
  d Why is an alloy of tin and lead used for solder, rather than pure tin or pure lead?
  e Below what temperature would an alloy containing equal quantities of tin and lead be solid?

% lead	100	90	80	70	60	50	40	30	20	10	0
% tin	0	10	20	30	40	50	60	70	80	90	100
melting point (°C)	327	304	280	257	234	211	188	194	209	223	237

Dental Fillings

Teeth have evolved over millions of years to become perfect biting and chewing machines. The outer layer is made of enamel, a dense, brittle material composed of the mineral calcium phosphate. Beneath the enamel is a layer of dentine. This is a bone-like substance, softer than enamel, and containing protein as well as calcium phosphate. Some of the properties of enamel and dentine are shown in the table below.

	compressive strength (MPa)	elastic modulus (GPa)	hardness (Knoop number)
enamel	900	47	343
dentine	300	9	68

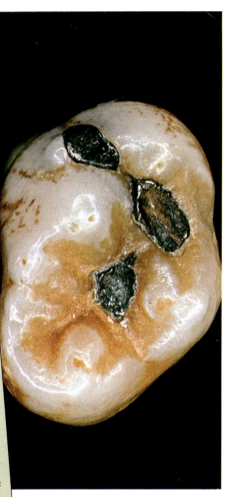

The compressive strength shows the maximum force that the material can take pressing down on it. The elastic modulus shows the amount by which the material can 'give'. The lower it is, the more the material can 'give'. When biting, the very hard enamel on the surface of the tooth is able to withstand very high compressive forces. Although it is brittle, the elasticity of the dentine beneath it cushions it, and makes it much less likely to break.

If a tooth decays the decayed part is removed and a filling is put in. The material used for the filling must have similar properties to the tooth. It must also be resistant to the warm, moist, salty, and sometimes acidic conditions inside the human mouth. Also, the material must be able to be moulded to fit the cavity exactly, bind tightly to the tooth and then harden quickly. Ideally, it should also look as 'tooth-like' as possible.

Several different filling materials have been developed. Perhaps the most familiar is **mercury amalgam**. This is an alloy of 12 % tin, 3 % copper, 0.2 % zinc, 35 % silver and 50 % mercury. It is mixed just before the filling is put in, and stays soft and pliable for long enough for the dentist to pack and mould it into the cavity. As it hardens, it expands slightly, which helps to hold it firmly in the tooth. The first material to be used for fillings was **gold**. This can be worked to make it soft enough to put into the tooth cavity, and it then hardens inside the mouth. However, pure gold always remains rather soft, and a gold-silver-copper alloy makes a better filling. Gold is a very unreactive metal so has high corrosion resistance. However, it is also very expensive. For fillings at the front of the mouth, appearance may be very important. Acrylic polymers have been developed which can match the colour of teeth perfectly, but these do not yet have such good mechanical properties as the metal alloys.

1 From what materials are enamel and dentine composed?

2 Describe the different properties of dentine and enamel. Explain how they enable a tooth to perform its function efficiently.

3 Why is it advantageous to have a layer of enamel on the outside of the dentine?

4 What properties are needed in a dental filling material?

5 What advantages does mercury amalgam have over gold, when used as a dental filling?

6 Acrylic polymers are still not widely used for fillings. Why is this?

12 IONIC COMPOUNDS

Atoms may gain or lose electrons, to become ions. Positive and negative ions are strongly attracted to each other.

Some atoms tend to lose electrons

You have seen in the last few pages that the atoms of metallic elements tend to lose electrons. Why is this?

Figures 12.1 and 12.2 show the structure of the atoms of sodium and neon. Neon has 10 protons and 10 electrons. The electrons are arranged in two complete shells. Neon is a very **stable** atom. It has a complete outer shell and no tendency to gain or lose electrons.

Sodium, however, has 11 protons and 11 electrons. It has two complete electron shells, and a single electron in its outer shell. This is a very **unstable** arrangement. The sodium atom has a strong tendency to lose this single electron, so that it ends up with a complete outer shell. When a sodium atom loses its outer electron, it becomes an **ion** (Figure 12.3).

An ion is an atom or group of atoms with an overall electrical charge. The sodium ion has a positive charge, because it still has its original 11 protons, but only 10 electrons. An ion with a positive charge is called a **cation**.

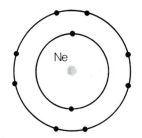

Fig. 12.1 A neon atom. The outer electron orbit is complete. This makes neon a very stable, unreactive element.

	charge
11 protons	= +11
11 electrons	= −11
total	= 0

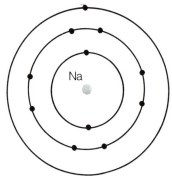

Fig. 12.2 A sodium atom. There is a single electron in the outer orbit. If the atom loses this electron, it will end up with a full outer orbit, like neon.

	charge
11 protons	= +11
10 electrons	= −10
total	= +1

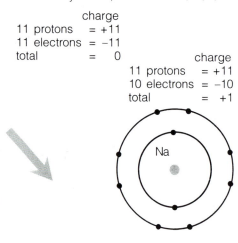

Fig. 12.3 A sodium ion. When a sodium atom loses its lone electron, it becomes a positive ion. It has a positive charge, because it now has one more proton than it has electrons.

Some atoms tend to gain electrons

Figures 12.4 and 12.5 show an argon atom and a chlorine atom. Argon, like neon, has a complete outer shell of electrons. It is a very stable atom, with no tendency to lose or gain electrons. Chlorine, however, has an outer shell which needs one more electron to complete it. So chlorine atoms have a strong tendency to gain electrons. When a chlorine atom gains an electron, it ends up with 17 protons and 18 electrons. This gives it an overall negative charge. It becomes a negatively charged ion, or an **anion**.

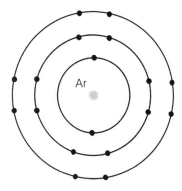

Fig. 12.4 An argon atom. Like neon, argon has a full outer electron orbit. This makes argon a very unreactive, stable element.

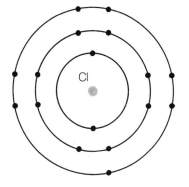

	charge
17 protons	= +17
17 electrons	= −17
total	= 0

Fig. 12.5 A chlorine atom. This has seven electrons in its outer orbit. It needs one more electron to fill up this orbit.

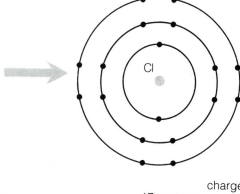

	charge
17 protons	= +17
18 electrons	= −18
total	= −1

Fig. 12.6 A chloride ion. This is a chlorine atom which has gained an electron to complete its outer orbit. It has a negative charge, because it has one more electron than it has protons.

Sodium and chlorine combine to make sodium chloride

What is likely to happen when sodium and chlorine atoms come together? A sodium atom can readily lose an electron, while a chlorine atom can readily gain one. It looks as though both of them can become stable if each sodium atom gives up an electron to a chlorine atom.

You can see what really happens if you watch sodium metal burning in chlorine. The sodium burns brightly and a white solid forms on the side of the jar. This solid is common salt. Its correct name is **sodium chloride**.

As sodium burns in chlorine, the sodium atoms lose electrons and the chlorine atoms gain them. The sodium atoms become positive ions, while the chlorine atoms become negative ions. The positive and negative ions are very strongly attracted to each other so they arrange themselves into a pattern like that shown in Figure 12.8. This is called a **lattice**. A grain of salt contains millions upon millions of sodium and chloride ions arranged like this. The force of attraction between the positive and negative ions is very strong. It is called an **ionic bond**.

Sodium chloride is a compound

You will remember that a substance which is made from just one kind of atom is called an **element**. Sodium and chlorine are both elements. Sodium chloride contains two elements, sodium and chlorine. A substance which is made of two or more kinds of atoms or ions that have joined together is called a **compound**. Sodium chloride is a compound. Because it is made of ions, it is called an **ionic compound**.

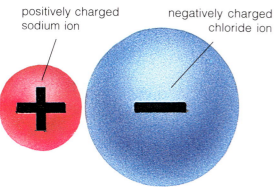

positively charged sodium ion

negatively charged chloride ion

Fig. 12.7 Sodium chloride. The positive and negative charges on the two ions hold them tightly together.

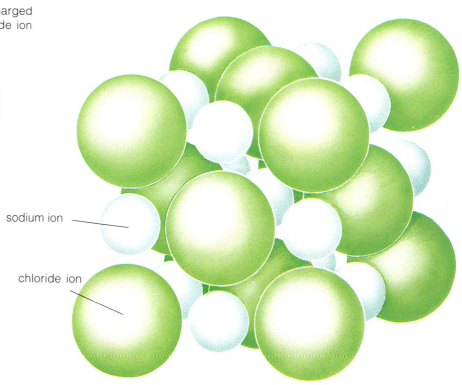

sodium ion

chloride ion

Fig. 12.8 A sodium chloride lattice. The sodium and chloride ions arrange themselves in a regular pattern, or lattice. Can you pick out the cubic shape? Salt crystals are cube-shaped because of this arrangement of ions.

A pair of ions is the smallest part of sodium chloride which can exist

Sodium ions and chloride ions combine to form the compound sodium chloride. The smallest part of a sodium chloride crystal which would still *be* sodium chloride would be one sodium ion and one chloride ion. It is unlikely that a single sodium ion and a single chloride ion would ever find themselves alone together! But it is possible, and this would be the smallest part of the compound which still has the properties of sodium chloride.

Questions

1 For each of the following, write a definition, and give an example:
a atom **b** ion **c** cation **d** anion
2 Use the information about electron configuration in Table 7.1 to predict whether each of these elements will form positive or negative ions:
a fluorine **b** potassium **c** lithium

13 USING FORMULAE

The name of a compound tells you which elements it contains. Formulae also tell you the relative numbers of the different kinds of atoms or ions in the compound.

The name of a compound often tells you which elements it contains

The name 'sodium chloride' tells you that this compound contains sodium and chlorine. When naming an ionic compound, the name of the positive ion always comes first. Its name is just the same as the name of the uncharged atom. We talk about 'sodium atoms' and 'sodium ions'.

The name of the negative ion comes last. You have probably realised that its name is not quite the same as the name of the uncharged atom. We talk about 'chlorine atoms' but 'chloride ions'. So the compound of sodium and chlorine is called sodium chloride. This rule almost always works. Table 13.1 gives you some more examples.

Some ions have a double charge

Figure 13.1 shows a magnesium atom. It has two electrons in its outer ring. If it loses these two electrons, it ends up with a complete outer ring. When it does this, it ends up with *two* more protons than electrons. It has a double positive charge. The formula for a magnesium ion shows this double charge. It is written **Mg²⁺**. Another common positive ion with a double charge is calcium, **Ca²⁺**.

Negative ions can also have double charges. Figure 13.3 shows an oxygen atom. It has six electrons in its outer shell and needs two more to complete it. When it gains them it becomes an oxide ion, **O²⁻**.

Question

If you had a tiny bit of NaCl containing only 1 million Na⁺ ions, how many Cl⁻ ions would it contain?

Elements	Positive ion	Negative ion	Name of compound
sodium, chlorine	sodium	chloride	sodium chloride
potassium, iodine	potassium	iodide	potassium iodide
magnesium, oxygen	magnesium	oxide	magnesium oxide
iron, sulphur	iron	sulphide	iron sulphide

Table 13.1   Some ionic compounds

The formula for sodium chloride is NaCl

The symbol for an atom of sodium is Na. When it loses an electron and becomes an ion, it has a positive charge. So the symbol for a sodium ion is **Na⁺**. The symbol for a chlorine atom is Cl. Chloride ions have a negative charge, so the symbol for a chloride ion is **Cl⁻**.

In sodium chloride there is one chloride ion for every sodium ion. The overall charge is nil, because the positive charges on the sodium ions balance the negative charges on the chloride ions. So the formula for sodium chloride is **NaCl**. The formula does not show the charges on the sodium or chloride ions, because when they are together the charges cancel each other out.

Every compound has a formula. The formula tells you a lot. It tells you what *kind* of atoms there are in the compound, and it tells you *how many* of each sort there are combined in that compound.

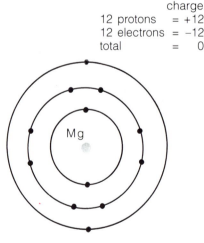

	charge
12 protons	= +12
12 electrons	= −12
total	= 0

Fig. 13.1   A magnesium atom. It has two electrons in its outer orbit.

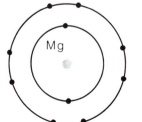

	charge
12 protons	= +12
10 electrons	= −10
total	= +2

Fig. 13.2   A magnesium ion, Mg²⁺. The two outer electrons are lost, leaving the ion with a complete outer orbit.

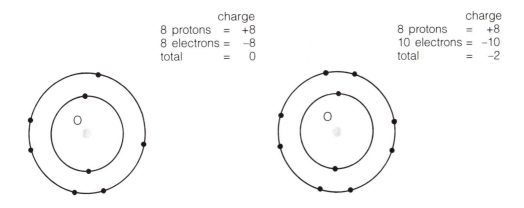

charge
8 protons = +8
8 electrons = −8
total   =   0

charge
8 protons   = +8
10 electrons = −10
total   = −2

Fig. 13.3 An oxygen atom

Fig. 13.4 An oxide ion, O^{2-}. An oxygen atom can achieve a full outer electron orbit by gaining two electrons.

The numbers of positive and negative ions in an ionic compound are not always equal

What happens if magnesium and chlorine react together? Magnesium atoms tend to lose electrons, becoming positive ions. Each magnesium atom loses two electrons, becoming Mg^{2+}. Chlorine atoms tend to gain electrons, becoming negative ions. Each chlorine atom gains one electron, becoming Cl^-.

So *one* magnesium atom can supply *two* chlorine atoms with the electrons they need. So when magnesium burns in chlorine, the atoms form a structure in which there is *one* magnesium ion for every *two* chloride ions.

The formula for magnesium chloride shows this. It is written **MgCl₂**. The small ₂ after the Cl tells you that there are two chloride ions in the compound magnesium chloride. There is no number after the Mg. This means that there is one Mg ion in the compound. Whenever no number is written after a symbol in a formula, this means that there is only one atom or ion of that element in a molecule. So $MgCl_2$ tells you that, in magnesium chloride, there are two chloride ions for every one magnesium ion.

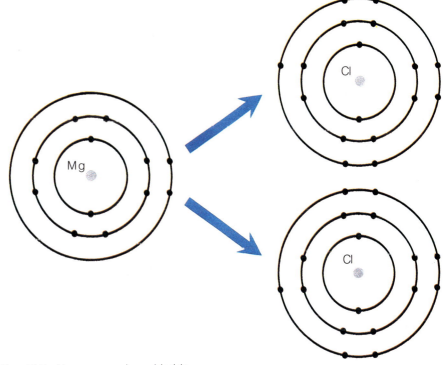

Fig. 13.5 How magnesium chloride forms. If one magnesium atom gives one electron to each of two chlorine atoms, then all three end up with complete outer electron orbits.

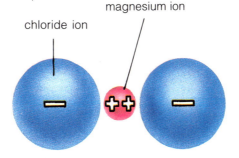

chloride ion

magnesium ion

Fig. 13.6 Magnesium chloride, $MgCl_2$. There is one magnesium ion for every two chloride ions.

Question

If you had a tiny bit of magnesium chloride, containing only 1 million Mg^{2+} ions, how many Cl^- ions would it contain?

14 TRANSITION METALS

The transition metals are the elements which fill the centre block in the Periodic Table. Most of them can form more than one kind of ion.

Most transition metals can form more than one kind of ion

The transition metals are the ones in the centre block of the Periodic Table (page 27). They include many common metals, such as iron and copper.

The atoms of transition metals, like all metals, can become more stable by losing electrons and forming positive ions. But most transition metals can become stable by losing **different numbers** of electrons. This means that they can form *more than one kind of ion*.

Copper is a typical example. The copper atom can become stable by losing either one or two electrons. If it loses one electron, it forms an ion with a single positive charge, **Cu$^+$**. This ion is called **copper (I)**. If it loses two electrons, it forms an ion with a double positive charge, **Cu^{2+}**. This ion is called **copper (II)**.

Figure 14.2 shows how these ions may combine with chloride ions. Each chloride ion has a single negative charge, Cl$^-$. Copper (I) ions have a single positive charge, so these two types of ion will combine together in equal numbers to form the ionic compound **CuCl**. This compound is called **copper (I) chloride**. Copper (II) ions, however, have a double positive charge, Cu^{2+}. So *two* chloride ions are needed to combine with each copper (II) ion. The ionic compound formed has the formula **CuCl$_2$**. It is called **copper (II) chloride**.

Fig. 14.1 The salts of transition elements are often coloured. The green powder is nickel (II) chloride; the bluer one is copper (II) chloride.

A copper (I) ion has a single positive charge

A chloride ion has a single negative charge

A copper (II) ion has a double positive charge

CuCl, copper (I) chloride. One copper (I) ion combined with one chloride ion balances their charges.

CuCl$_2$, copper (II) chloride. One copper (II) ion combined with two chloride ions balances their charges.

Fig. 14.2 Copper (I) chloride and copper (II) chloride

Transition metals form coloured compounds

Most of the compounds of transition elements are coloured. The colours are often different for the different kinds of ions formed by one metal. Copper (II) compounds, for example, are often blue, while copper (I) compounds are often white or green.

Iron is another important transition metal. It forms two sorts of ions, Fe^{2+} and Fe^{3+}. These are called **iron (II)** and **iron (III)**. Iron (II) compounds are usually pale green, while iron (III) compounds are usually yellow or brown.

Table 14.1   Some elements which form ions.

Element	Symbol	Ion	Name of ion
Potassium	K	K^+	Potassium
Sodium	Na	Na^+	Sodium
Lithium	Li	Li^+	Lithium
Magnesium	Mg	Mg^{2+}	Magnesium
Calcium	Ca	Ca^{2+}	Calcium
Iron	Fe	Fe^{2+}	Iron (II)
		Fe^{3+}	Iron (III)
Copper	Cu	Cu^+	Copper (I)
		Cu^{2+}	Copper (II)
Zinc	Zn	Zn^{2+}	Zinc
Aluminium	Al	Al^{3+}	Aluminium
Lead	Pb	Pb^{2+}	Lead
Fluorine	F	F^-	Fluoride
Chlorine	Cl	Cl^-	Chloride
Bromine	Br	Br^-	Bromide
Iodine	I	I^-	Iodide
Oxygen	O	O^{2-}	Oxide
Sulphur	S	S^{2-}	Sulphide

Questions

1 a What is the symbol for a sodium atom?
b What is the symbol for a sodium ion?
c Does a sodium atom lose or gain electrons when it becomes an ion?
d How many electrons does it lose or gain?

2 a What is the symbol for a magnesium ion?
b How many electrons does a magnesium atom lose when it becomes an ion?
c What is the symbol for an oxygen ion?
d How many electrons does an oxygen atom gain when it becomes an ion?
e How many magnesium atoms will be needed to supply the electrons for one oxygen atom when they become ions?
f The compound formed when magnesium ions and oxygen ions combine together is called magnesium oxide. Write down the formula for magnesium oxide.

3 a How many electrons does a chlorine atom gain when it becomes a chloride ion?
b How many electrons does a calcium atom lose when it becomes an ion?
c How many chlorine atoms can be supplied with electrons by one calcium atom?
d Write down the formula for calcium chloride.

4 a Potassium atoms lose one electron when they become ions. Write down the symbol for a potassium ion.
b Sulphur atoms gain two electrons when they become ions. Write down the symbol for a sulphur ion.
c How many potassium atoms will be needed to supply the electrons for one sulphur ion?
d Write down the formula for potassium sulphide.

5 Explain why:
a Sodium atoms form positive ions, but chlorine atoms form negative ions.
b A potassium ion has a charge of $+$, but a magnesium ion has a charge of $2+$.
c The ions in sodium chloride are held closely together.
d A sodium ion is written as Na^+, and a chlorine ion is written as Cl^-, but the ionic compound sodium chloride is written as NaCl.

6 Name the following compounds (refer to the Periodic Table if you need to):
a KCl **b** MgO **c** Li_2S **d** CaF_2 **e** Na_2S
f $PbCl_2$ **g** SrO **h** RbF

7 The formula for aluminium fluoride is AlF_3.
a How many fluoride ions are there for every aluminium ion?
b How many electrons does one fluorine atom gain when it becomes an ion?
c From your answers to parts **a** and **b**, work out how many electrons one aluminium atom loses when it becomes an ion.
d Write down the symbol for an aluminium ion.
e What is the symbol for an oxide ion?
f If powdered aluminium is heated strongly in oxygen, the aluminium and oxygen atoms can react together to form aluminium oxide. The formula for aluminium oxide is Al_2O_3. Explain clearly why the aluminium and oxide ions combine together in these proportions. You may like to include a diagram in your answer.

8 Work out and write down formulae for these ionic compounds:
a calcium chloride
b calcium oxide
c potassium sulphide
d lithium oxide
e lithium bromide
f magnesium fluoride

9 Name these compounds of transition metals:
a Cu_2O **b** CuO **c** FeO **d** Fe_2O_3
e $CuCl_2$ **f** $FeCl_3$

10 Write formulae for:
a iron (II) chloride
b copper (I) sulphide
c copper (II) chloride
d copper (I) iodide
e iron (II) sulphide

11 Use the Periodic Table on page 27 to explain the following:
a All elements in Group 1 form an ion with one positive charge.
b All elements in Group 7 form an ion with one negative charge.

————— *E X T E N S I O N* —

15 COVALENT BONDS

Atoms can often share electrons to achieve a stable electron structure. This sharing holds the atoms closely together, and is called a covalent bond.

Atoms can fill their outer shells by sharing electrons with other atoms

You have seen that atoms tend to lose or gain electrons in order to end up with full outer electron orbits. If they completely lose or gain an electron, they become ions. Positive and negative ions are strongly attracted together, forming ionic bonds between themselves. Compounds formed in this way are called ionic compounds.

But how do compounds form when neither type of atom in the compound can lose electrons and form a positive ion? For example, bromine and chlorine form a compound with the formula BrCl. Yet both bromine atoms and chlorine atoms form ions by gaining electrons. They both form ions with a negative charge. If they are both gaining electrons, where can they be getting them from? Anyway, as both bromide and chloride ions are negatively charged, they would push each other apart – because if two particles are *both* negatively charged (or both positively charged) they repel each other. Clearly, BrCl cannot be an ionic compound. There must be another way that atoms can be held together.

In ionic compounds atoms fill their outer orbits by losing or gaining electrons. But there is another way in which they can do this. Atoms can fill their outer orbits by **sharing** electrons with other atoms.

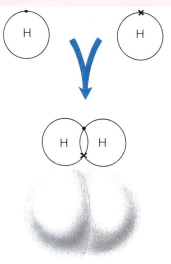

Fig. 15.1   A hydrogen molecule, H₂. Two hydrogen atoms can achieve complete outer electron shells by sharing electrons.

Hydrogen atoms pair up and share electrons

Figure 15.1 shows how hydrogen atoms share electrons with each other. Hydrogen atoms only have one electron and one proton. To fill their electron orbit, they need one more electron. If two hydrogen atoms share their single electrons with each other then each has, in effect, a full electron orbit. The two hydrogen atoms are held closely together when they share electrons like this. The bond between them is called a **covalent bond**.

The pair of bonded hydrogen atoms is a hydrogen **molecule**. It is the smallest part of the element hydrogen which can exist on its own. The formula for a hydrogen molecule is **H₂**, showing that the hydrogen atoms are linked in pairs.

Fig. 15.2   A chlorine molecule, Cl₂. One pair of electrons is shared.

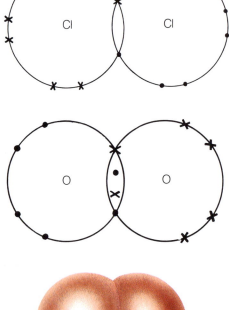

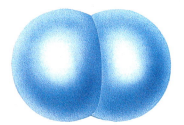

Fig. 15.3   A nitrogen molecule, N₂. Three pairs of electrons are shared between the two nitrogen atoms, so the bond between them is a triple bond.

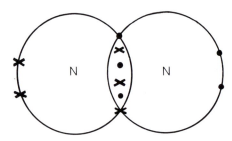

Fig. 15.4   An oxygen molecule, O₂. Two pairs of electrons are shared, so the bond between the oxygen atoms is a double bond.

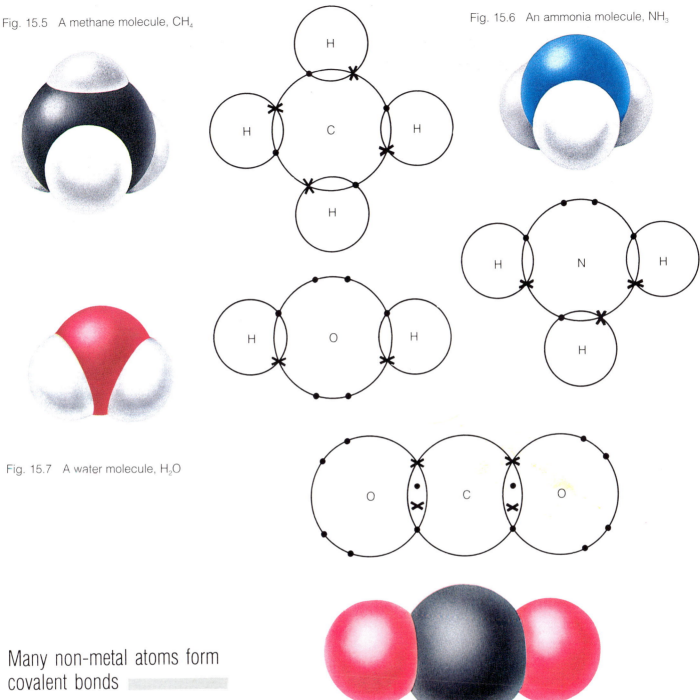

Fig. 15.5   A methane molecule, CH₄

Fig. 15.6   An ammonia molecule, NH₃

Fig. 15.7   A water molecule, H₂O

Fig. 15.8   A carbon dioxide molecule, CO₂

Many non-metal atoms form covalent bonds

Figures 15.1 to 15.4 show some other examples of atoms which form covalent bonds. Sometimes they do this with other atoms like themselves. Oxygen gas, for example, is made up of pairs of oxygen atoms linked with covalent bonds. Its formula is O_2. Chlorine gas is also made of pairs of chlorine atoms, forming chlorine molecules, Cl_2.

Atoms can also form covalent bonds with different sorts of atoms. When they do this, a **covalent compound** is formed. Figures 15.5 to 15.8 show some examples of covalent compounds. **Water, ammonia, carbon dioxide** and **methane** are four very important covalent compounds.

One pair of shared electrons is one covalent bond

You can see from the diagrams of the covalent compounds that they share their electrons in pairs. Each *pair* of electrons shared is *one* covalent bond. Hydrogen molecules, for example, have one bond, because they share one pair of electrons.

But oxygen atoms share *two* pairs of electrons when they form oxygen molecules. So they have *two* bonds between them. This arrangement is often called a **double bond**.

Nitrogen atoms share even more electrons. Each nitrogen atom has five electrons in its outer shell, and needs three more to make a stable outer orbit of eight electrons. So two nitrogen atoms share *three* pairs of electrons to make nitrogen molecules, N_2. This is sometimes called a **triple bond**.

16 MORE ABOUT COVALENT BONDS

There is an enormous variety of covalent compounds.

What are you made of?

Did you know that living things are made almost entirely of six elements? Apart from small traces of other elements, you consist almost entirely of compounds made from carbon, hydrogen, nitrogen, oxygen, phosphorus, and sulphur. How can something as complex as a living animal be made from such a simple set of substances? The answer is that these six elements can produce an almost endless variety of covalent compounds. The diagrams on this page show three more covalently bonded substances made from the same elements as those shown in Topic 15.

Two elements can join in different ways to make different compounds

Figure 16.1 is a diagram of the bonding in a molecule of **hydrogen peroxide**. The molecule is made of hydrogen atoms and oxygen atoms, just like the water molecule on the previous page. But while a water molecule consists of two hydrogen atoms and one oxygen atom and so has the formula H_2O, a hydrogen peroxide molecule consists of two hydrogen atoms and two oxygen atoms and so has the formula H_2O_2.

Although they are made of the same elements, hydrogen peroxide and water are different compounds. They behave differently, or we could say they have **different properties**. Note that in the diagram of hydrogen peroxide every atom shares electrons and achieves a full outer orbit. Note also that each atom started with its correct number of outer electrons – one in the case of hydrogen and six in the case of oxygen.

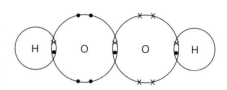

Figure 16.1 Hydrogen peroxide molecule H_2O_2

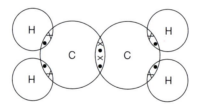

Figure 16.3 Ethene molecule C_2H_4

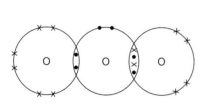

Figure 16.2 Ozone molecule O_3

Atoms of a single element can also bond in different ways

Having the atoms bonded in a different way can also give us different forms of elements. Figure 16.2 shows three oxygen atoms bonded together. This substance is called **ozone** and has the formula O_3. Note that it is an element – each molecule is made of one type of atom only. This form of oxygen is not very common in the air we breathe at ground level but there is a layer of it high in the atmosphere. The bonding diagram looks a bit strange when examined closely. The bond on the left is made by two electrons both from the same oxygen atom, but the rules are still not broken – each oxygen atom begins with six outer electrons and achieves a share in eight.

Question

1 Figure 16.3 shows a molecule of ethene. Look at the diagram and answer the following questions.
a How many atoms are there in an ethene molecule?
b Which elements are present in ethene?
c Which compound on the previous page was made from these elements?
d Will these two compounds made from the same elements have exactly the same properties?
e What is the formula of ethene?
f How many outer electrons did each atom in the ethene molecule have before it bonded?
g How many electrons does bonding give each atom a share in?
h Where in this molecule are there single covalent bonds?
i Where in this molecule is there a double covalent bond?

Some advice on drawing covalent bonding diagrams

Sometimes you may be asked to draw diagrams to show the structure of a molecule of a covalent compound. If you follow these rules, you are more likely to get it right!

1 Check that every atom has a filled outer orbit. In this diagram of 'ammonia' the hydrogen atoms do not – so it must be wrong.

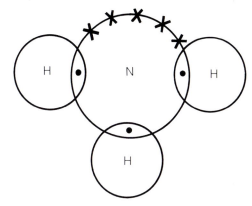

2 Check that each atom has started off with the correct number of electrons. Dot and cross diagrams help you to check this. In this diagram of 'ammonia' the nitrogen has begun with six outer electrons, but it should have five.

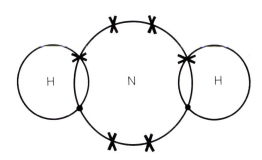

3 Even if you cannot see the solution immediately, follow rules **1** and **2** and try – it is surprising how much easier it can become!

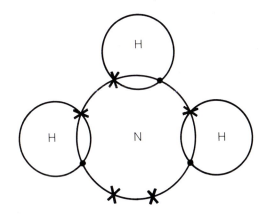

Questions

1 From the molecules shown in Topic 15, name:
a two elements
b two compounds
c one element which exists as a gas at room temperature
d one compound which exists as a liquid at room temperature
e one covalent element whose atoms are held together by double bonds
f one compound where the molecules contain two different sorts of atoms
g the compound with the formula NH_3
h one compound which has four bonds

2 Below are the atomic numbers of six elements:

Element	Atomic number
H	1
Cl	17
C	6
N	7
O	8
S	16

a Draw electronic structures for the atoms of these six elements.

b Draw diagrams to show how electrons are shared to make molecules of the following covalent compounds (show outer electrons only):

i HCl
ii CCl_4
iii NCl_3
iv OCl_2

c Now try the same for these molecules – they are quite a bit harder!

i CO (hint – unequal sharing)
ii S_8 (hint – ring-shaped)
iii C_2H_2 (hint – all four atoms are in a straight line).

EXTENSION

Ionic and covalent compounds tend to behave differently from each other.

INVESTIGATION 17.1

Comparing ionic and covalent compounds

You are going to do a set of experiments on six compounds. A, B and C are ionic compounds. D, E and F are covalent compounds. When you have completed the investigation, you should be able to make generalisations about the properties of ionic and covalent compounds.

1 Read through the investigation. Then design a results chart which you can use as you perform the experiments. Ask for help if you are not sure how to do this.

2 WEAR SAFETY SPECTACLES. Place a little of substance A on a piece of ceramic paper. Heat in a blue Bunsen burner flame using tongs. Does it melt or burn easily? Observe what happens, and fill in the results in your chart. Now do the same for the other five compounds.

3 Put some water in a test tube. Add a little of substance A and shake. Does A dissolve in water? Fill in your result in the chart.

4 If A does dissolve, use the circuit provided to see if the solution conducts electricity. Fill in your result in the chart.

5 Repeat steps **3** and **4** for the other five substances.

6 Repeat steps **3** and **4**, but use 1,1,1-trichloroethane instead of water. This is a liquid which does not contain water. (Your teacher may demonstrate this part of the investigation for you.)

Questions

It is not always possible to draw conclusions about behaviour which are true for all ionic and covalent compounds. Some of your results may not fit neatly into a pattern. But you should be able to see enough of a pattern to answer the following questions.

1 Which are easier to melt – ionic or covalent substances?
2 Are ionic substances more likely to dissolve in water, or·in a non-aqueous solvent (one which does not contain water)?
3 Are covalent substances more likely to dissolve in water, or in a non-aqueous solvent?
4 Do solutions of ionic substances conduct electricity?
5 Do solutions of covalent substances conduct electricity?

Question

Fig. 17.1 Three of these aqueous solutions are of covalent compounds, and three of ionic compounds. Which are which? Why? What is slightly unusual about the three covalent compounds?

U

V

W

X

Y

Z
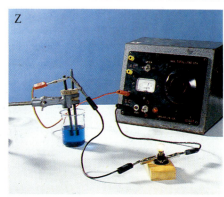

An explanation of the properties of ionic and covalent compounds

MELTING POINTS

In an ionic compound all the ions in the lattice are held firmly together. The bonds are strong. If the compound is heated the heat energy makes the particles vibrate. But it takes a lot of heat energy to break the particles completely apart. This means that ionic compounds need a lot of heat to change them from solids to liquids. In other words they have **high melting points**. All ionic compounds are solids at room temperature.

In a covalent substance the atoms in each molecule are held firmly together by covalent bonds. But the individual molecules are held together by only weak forces between them. This means that not very much heat is needed to separate the molecules. Quite low temperatures are enough to change the substance from a solid to a liquid or gas. In other words covalent substances have **low melting** and **boiling points**. Many covalent substances are gases at room temperature.

SOLUBILITY IN WATER

In an ionic compound, the charges on the ions attract water molecules to them. This means that they will **dissolve in water**. You will find more about how substances dissolve in water on page 63.

In a covalent substance the molecules often do not attract water molecules as much. This means that they often **do not dissolve in water**, but they can **dissolve in non-aqueous solvents**.

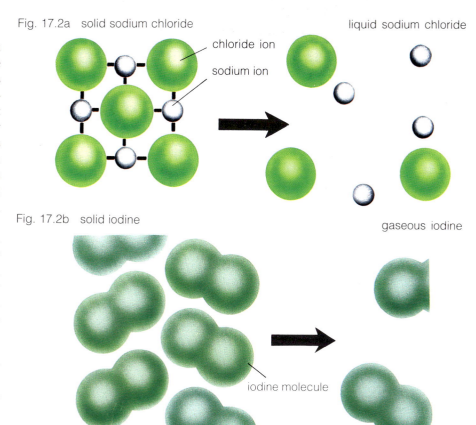

Fig. 17.2a   solid sodium chloride   liquid sodium chloride

chloride ion

sodium ion

Fig. 17.2b   solid iodine   gaseous iodine

iodine molecule

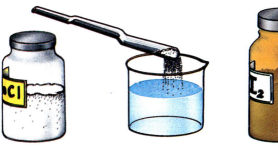

Fig. 17.3a   Sodium chloride readily dissolves in water.

Fig. 17.3b   Iodine hardly dissolves at all in water.

CONDUCTION OF ELECTRICITY

An electric current is a flow of charged particles. A solution of an ionic compound has freely-moving charged particles. This means it **can easily conduct electricity**.

Covalent substances **cannot conduct electricity**.

Fig. 17.4a   A solution of sodium chloride in water conducts electricity.

Fig. 17.4b   A solution of iodine in petrol does not conduct electricity.

18 LARGE MOLECULES

Some covalent compounds have molecules that are extremely large.

The atoms of some substances form giant molecules

Some substances don't seem to have any of the three types of structure you have met so far. They are made of non-metals, so they are not metallic or ionic. They have extremely high melting and boiling points, so they can't be covalent compounds with small molecules. The atoms in these substances are held together by covalent bonds, but they are not held in small groups. Instead, the atoms are held together in vast arrays to form giant molecules.

Diamond is one example of a substance with a giant molecular structure. Every atom is the same. They are all carbon atoms, so diamond is one form of the element carbon. Figure 18.1 shows a small part of a diamond crystal. Each atom is held to the four atoms around it by strong covalent bonds. There are no weak points in this structure. That is

why diamonds are so hard. Since a lot of energy is required to break each diamond atom away from all those around it, diamond has a very high melting point.

Graphite also has a giant molecular structure (Figure 18.2). Once again every atom is a carbon atom, so graphite is another form of the element carbon. Graphite is used for pencil lead, so pencil 'lead' and diamonds are made from different forms of the same substance. In graphite each carbon atom is bonded to three other carbon atoms around it by strong covalent bonds to form layers of carbon atoms. Like diamond this means graphite has a very high melting point. But the layers are not strongly bonded to each other, so they can slide over each other. That is what is happening when you use a

pencil – layers of graphite atoms slide off the end of your pencil onto the paper, leaving a black line on the paper. The structure of graphite allows electrons to move along the layers, so graphite conducts electricity, which is very unusual for a non-metal.

Figure 18.3 shows a third example of a giant covalent structure. This is a compound, silicon dioxide, familiar to us as sand and quartz. As with the two previous examples the diagram can show only a few atoms to give some impression of the structure, but even a single grain of sand contains billions of billions of atoms bonded in this way! Since covalent bonds are strong, silicon dioxide is hard and has a high melting point.

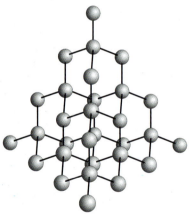

Fig. 18.1 The structure of diamond

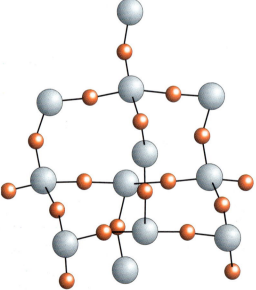

Fig. 18.3 The structure of silicon dioxide

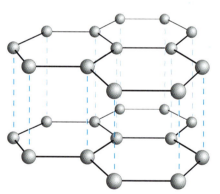

Fig. 18.2 The structure of graphite

All substances containing atoms that are held together by covalent bonds are known as covalent substances, but we can now sub-divide this category further. In some of these substances the bonds hold the atoms in vast arrays or lattices. These substances are said to have a **giant covalent** or **giant molecular** structure. Diamond and the other substances shown on page 48 are examples of giant covalent or giant molecular structures. In other covalent substances the bonds hold the atoms in small (or comparatively small) groups. These substances are said to have **simple molecular** structures. Water, ammonia, and the other substances shown in Topics 15 and 16 are examples of simple molecular structures. Some substances with simple molecular structures can have quite big molecules, for example the DNA molecule in Figure 18.5, but even a molecule consisting of thousands of atoms is tiny compared to a true giant molecule like silicon dioxide or graphite.

Question

1 Find out:
  a more about why graphite conducts electricity
  b about another form of carbon called buckminsterfullerene
  c why diamond is denser than graphite
  d a more accurate estimate of the number of atoms in a grain of sand.

Many biological molecules are very big

You will have found when you did Investigation 17.1, that not all substances fit neatly into one of these two groups. In particular, there are many covalent substances which have some properties more like those of ionic compounds.

Very big covalent molecules tend to be solid at room temperature, and have high melting and boiling points. Sugar is a good example. The large mass of the sugar molecules means that a lot of heat energy is needed to get them moving freely.

Some covalent compounds are soluble in water. This is because they do actually have small charges on them. These charges are not as large as those on ions. They are called **dipoles**, and you will find more about them on page 59. Sugar molecules, for example, have small positive and negative charges on them, so they will dissolve in water.

All the big biological molecules – proteins, carbohydrates, and fats – are covalent. But because they are big molecules, they are often solid at room temperature. Also many of them have dipoles, so they will dissolve in water.

— E X T E N S I O N —

Fig. 18.5 Many biological molecules contain thousands of atoms. This is a computer graphic representation of a tiny part of a DNA molecule. Most of the atoms in the molecule are bonded covalently. DNA is the genetic material found in all living cells.

Fig. 18.4 A dog's fur is made of keratin, which is a protein. Proteins are covalent compounds consisting of big molecules. Keratin is insoluble in water, but many proteins, such as haemoglobin, are soluble in water.

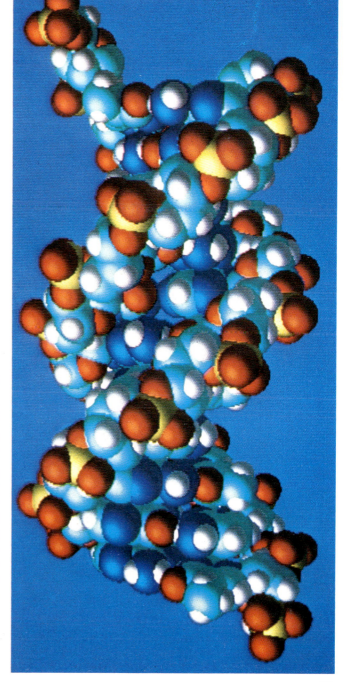

19 COMPOUND IONS

Ions can be formed from groups of atoms.
These groups are called compound ions.

A compound ion is an ion formed from a group of atoms

You may remember that the definition of an ion is **an atom or group of atoms with an overall positive or negative charge**. Ions have charges because the number of protons and electrons is unequal. A positive ion has more protons than electrons. A negative ion has more electrons than protons. So far, all the ions you have met in this book have been formed from single atoms. But there are many important ions which are formed from **groups** of atoms. Some of them are shown in Figures 19.1 to 19.5. The atoms in the group are joined together by sharing electrons – in other words, they are held together by covalent bonds. But, even by sharing, every atom in the group doesn't quite get a complete outer electron shell. The group of atoms still needs to gain or lose an electron or two in order to end up with all the outer shells completely filled.

Look at the diagram of the **carbonate ion**. The three oxygen atoms fill their outer shells by sharing electrons with the carbon atom. But for *all* the atoms in the group to end up with complete outer shells, the group still needs to gain two more electrons. When it does this, the total number of electrons in the group is two more than the total number of protons. So the group ends up with a double negative charge. It becomes a **negative ion**. The carbonate ion is one example of a compound ion. A **compound ion** is a group of atoms with an overall positive or negative charge.

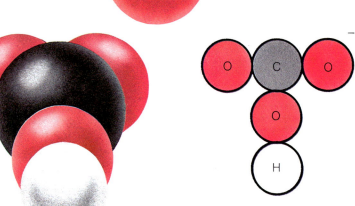

Fig. 19.1   A carbonate ion, CO_3^{2-}

Fig. 19.2   A hydrogencarbonate ion, HCO_3^-

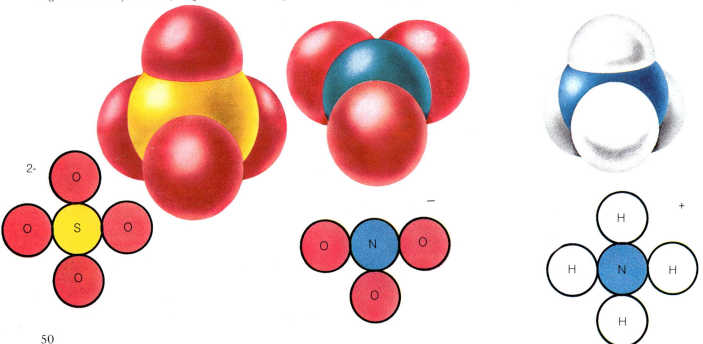

Fig. 19.3   A sulphate ion, SO_4^{2-}

Fig. 19.4   A nitrate ion, NO_3^-

Fig. 19.5   An ammonium ion, NH_4^+

Compound ions form compounds in the same way as simple ions

Compound ions combine with other ions to form ionic compounds. You can work out their formulae in the same way as before. For example, the carbonate ion, CO_3^{2-}, has a double negative charge. A sodium ion, Na^+, has a single positive charge. So if a lattice is formed with two sodium ions for every one carbonate ion, the charges will cancel out. The formula for this ionic compound is **Na_2CO_3**. Its name is **sodium carbonate**.

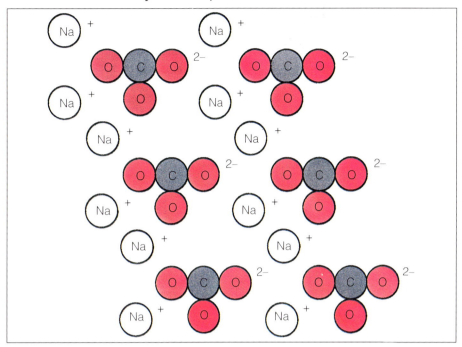

Fig. 19.6 Sodium carbonate, Na_2CO_3. The Na^+ and CO_3^{2-} ions form a lattice, in which there are two sodium ions for every carbonate ion. The ions are held strongly together by the attraction between the positive and negative charges.

Brackets are needed to show more than one compound ion in a formula

Most of the common compound ions have negative charges, but the **ammonium** ion, **NH_4^+**, has a single positive charge. In ammonium chloride, since the chloride ion has a single negative charge, one ammonium ion is needed to balance the charge on one chloride ion. The formula is written as **NH_4Cl**. But in ammonium sulphate, the sulphate ion has a double negative charge. This means that *two* ammonium ions are needed to balance the charge on one sulphate ion.

How can you write a formula showing that there are two NH_4^+ ions in a molecule? You could write NH_4NH_4, but that is rather untidy. You certainly could not write NH_{42} because that would mean that there were 42 hydrogen atoms in the molecule! The only way to show two ammonium ions is to write **$(NH_4^+)_2$**. The bracket shows that the $_2$ applies to everything inside the bracket. So the formula for ammonium sulphate is **$(NH_4^+)_2SO_4$**.

In **magnesium nitrate**, the magnesium ion has a double charge, Mg^{2+}. The nitrate ion has a single charge, NO_3^-. So two nitrate ions are needed to balance the charge on each magnesium ion. The formula for magnesium nitrate is written as **$Mg(NO_3)_2$**. The bracket is needed to show that the $_2$ applies to the whole of the nitrate ion, NO_3^-.

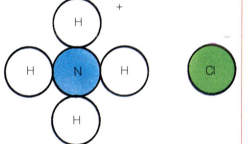

Fig. 19.7 Ammonium chloride, NH_4Cl

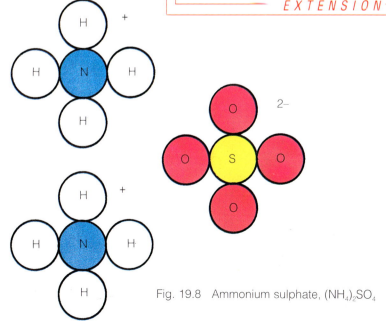

Fig. 19.8 Ammonium sulphate, $(NH_4)_2SO_4$

Questions

1 Name these compounds:
   a NH_4Cl **b** $Ca(NO_3)_2$ **c** $MgSO_4$
   d $NaOH$

EXTENSION

EXTENSION

51

Questions

1 Some properties of Group 1 elements are given in the table.

name	symbol	mass number	melting point /°C	reaction with water
lithium	Li	7		produces a steady stream of bubbles
sodium	Na		98	whizzes around the surface
	K	39	64	catches fire on contact
rubidium		85	39	explodes on contact
caesium	Cs	133	29	

a Complete the table to show:
  • your prediction for the melting point of lithium
  • the mass number of sodium
  • the name of the third element in the table
  • the symbol for rubidium
  • your prediction for the reaction of caesium with water

b Two of these elements are not kept in schools. Suggest and explain which two these are.

c The three of these elements which are kept in schools are stored in oil. Suggest why this is done.

2 The table shows some properties of some common elements.

element	metal	conducts electricity	conducts heat
copper		yes	yes
hydrogen	no		no
zinc	yes		yes
sulphur		no	no
phosphorus	no	no	
sodium	yes	yes	

Copy and complete the table.

3 a Explain how covalent bonds form in a molecule.

b Compare the properties of ionic and covalent compounds, using named examples wherever possible in your answer, and giving explanations for the differences between them.

4 Suggest explanations for each of the following, in terms of the ionic or molecular structures of the substances concerned.

a Plugs for electric appliances are often made of plastic, but never of metal.

b The fertiliser ammonium nitrate should not be applied to soil when it is about to rain.

c A burn from melted sugar is much more damaging than a burn from hot water.

d Hydrogen sulphide, which is a covalent compound, is a gas at room temperature, while iron sulphide is a solid.

5 The diagram shows the electron arrangement in a molecule of ethanol.

$$
\begin{array}{ccccc}
& \mathbf{H} & \mathbf{H} & & \\
& {}_{\circ\times} & {}_{\bullet\times} & {}_{++} & \\
\mathbf{H} {}_{\times}^{\circ} & \mathbf{C} {}_{\circ\times}^{\bullet} & \mathbf{C} {}_{\bullet\times}^{\bullet} {}_{+}^{+} & \mathbf{O} {}_{++}^{+} {}_{\times} & \mathbf{H} \\
& \mathbf{H} & \mathbf{H} & & \\
\end{array}
$$

a How many atoms are there in this molecule?

b What is the molecular formula for ethanol?

c What do each of the four types of symbol (circles and crosses) represent?

d How many bonds does each carbon atom have?

e Which type of bonding is present in this molecule?

f Would you expect ethanol:
  • to be a solid or a liquid at room temperature?
  • to conduct electricity?

AIR AND WATER

The Earth is surrounded by a layer of gas which we call air. Air is a mixture of several different gases.

Air is a mixture of gases

The Earth's atmosphere contains several different gases. Most of it – 78 % – is **nitrogen**. Almost all of the rest – 21 % – is **oxygen**. The rest is made up of very small amounts of **carbon dioxide**, **argon** and other **noble gases**, and **water vapour**.

Table 20.1 Gases in the air

gas	percentage
nitrogen	78
oxygen	21
argon	1 (just under)
carbon dioxide	0.04
neon, krypton, xenon, water vapour	very small amounts

Carbon dioxide is needed for photosynthesis

Only a tiny proportion of the air – about 0.04 % – is carbon dioxide. Yet without this gas, there would be no plants or animals on Earth! Plants use carbon dioxide to make food, in the process of photosynthesis. All the food which animals eat is made from carbon dioxide from the air.

Carbon dioxide is used to make fizzy drinks. It is also used in some types of fire extinguishers. Because it is a heavy gas, it forms a 'blanket' over the flames, stopping oxygen getting in.

> You can test a gas to see if it is carbon dioxide by using limewater. Limewater is a solution of calcium hydroxide in water. When carbon dioxide is mixed with it, calcium carbonate (chalk) is formed. This is not soluble in water, so it makes the limewater look cloudy.

Fig. 20.1 The Earth's atmosphere. The atmosphere gets thinner and thinner as you go upwards. 80% of the atmosphere is in the lowest layer, the troposphere.

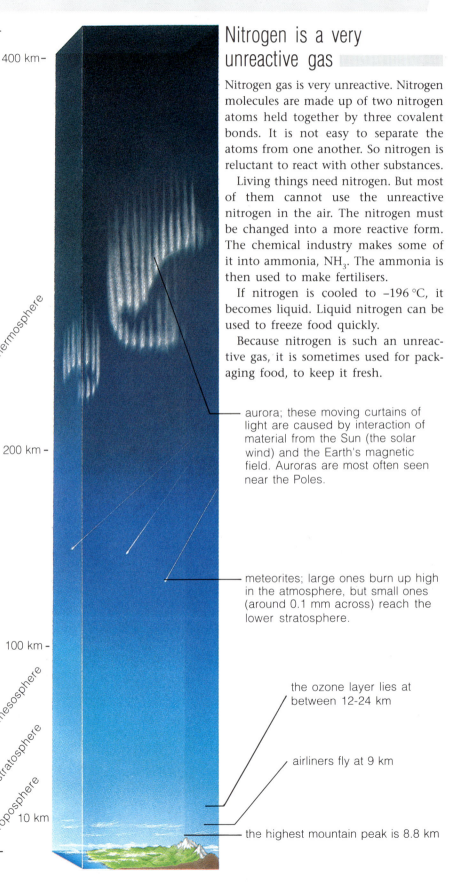

400 km –

200 km –

100 km –

10 km

thermosphere

mesosphere

stratosphere

troposphere

Nitrogen is a very unreactive gas

Nitrogen gas is very unreactive. Nitrogen molecules are made up of two nitrogen atoms held together by three covalent bonds. It is not easy to separate the atoms from one another. So nitrogen is reluctant to react with other substances.

Living things need nitrogen. But most of them cannot use the unreactive nitrogen in the air. The nitrogen must be changed into a more reactive form. The chemical industry makes some of it into ammonia, NH_3. The ammonia is then used to make fertilisers.

If nitrogen is cooled to $-196\,°C$, it becomes liquid. Liquid nitrogen can be used to freeze food quickly.

Because nitrogen is such an unreactive gas, it is sometimes used for packaging food, to keep it fresh.

aurora; these moving curtains of light are caused by interaction of material from the Sun (the solar wind) and the Earth's magnetic field. Auroras are most often seen near the Poles.

meteorites; large ones burn up high in the atmosphere, but small ones (around 0.1 mm across) reach the lower stratosphere.

the ozone layer lies at between 12-24 km

airliners fly at 9 km

the highest mountain peak is 8.8 km

Oxygen is needed for burning and respiration

Like nitrogen, oxygen gas is made up of molecules in which two atoms are joined together with covalent bonds. But oxygen is a reactive gas. Oxygen reacts easily with many substances to produce oxides. This can be a nuisance. For example, when it reacts with iron it forms iron oxide (rust).

Oxygen is needed for burning. When something burns, it combines with oxygen in the air and releases heat energy. Respiration inside living organisms is a similar process. Oxygen reacts with food substances and releases energy. This is why most organisms need to take in oxygen when breathing.

Substances burn more readily in pure oxygen than they do in air. This fact is used in the test for oxygen. Light a wooden splint, and blow it out. Then put it into the gas you think might be oxygen. If it is, the splint will burst into flame again.

INVESTIGATION 20.1

Measuring the percentage of oxygen in air

When copper is heated in air, it reacts with the oxygen in it, forming copper oxide. If you use plenty of copper, then *all* the oxygen in a sample of air will react. By comparing how much air you had to start with, and how much you have left when all the oxygen has been used up, you can find out how much oxygen there was in your sample.

1 Set up the apparatus as in the diagram. It is really important that everything fits tightly, with no air gaps. Notice that one syringe has its plunger pushed right to the end, so that it contains no air at all. The other one contains 100 cm³ of air.

2 Heat the copper wire strongly. As you heat it, push the plungers of the two syringes back and forth several times. The copper will gradually turn black, as it combines with the oxygen in the air to form copper oxide.

3 When you think that all the oxygen has had a chance to react with the copper, stop heating and let the apparatus cool down to room temperature again.
Then push all the air into one syringe, and measure its volume. It should be less than you started with. Write down the volume.

4 To make sure that all the oxygen really has been used up, heat the copper again, and push the syringes back and forth as before. Cool, and measure the volume of air. If it is the same as last time, then all the oxygen has gone. If not, you must repeat instructions 2 and 3.

5 Work out how much oxygen there was in your sample of air.

6 You had 100 cm³ of air to start with. Work out the percentage of air which was oxygen.

Questions

1 Why is it best to use copper wire or turnings in this experiment, rather than lumps of copper?

2 Why is it important for the apparatus to be airtight?

3 Why should you push the syringes back and forth while heating the copper?

4 Why must you wait for the apparatus to cool down before measuring the new volume of air?

5 What gases will be present in the air inside the syringe at the end of the experiment?

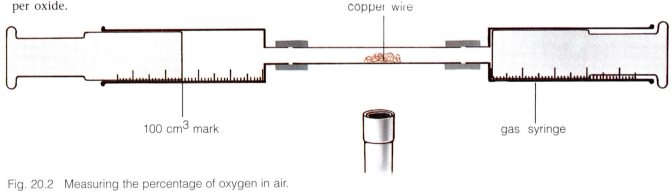

100 cm³ mark   copper wire   gas syringe

Fig. 20.2 Measuring the percentage of oxygen in air.

Questions

1 List the gases present in the air, giving the percentages of each.

2 a Give two uses of nitrogen.
b What is the boiling point of nitrogen?
c Why can living organisms not use nitrogen gas from the air?

3 a Give two uses of oxygen.
b What is the molecular formula for oxygen gas?
c Give two ways in which the chemical reactions involved in burning and respiration are similar.
d How can you test for oxygen?

4 a Why is carbon dioxide a very important gas for living organisms?
b How can you test for carbon dioxide?

21 THE CARBON CYCLE

Carbon dioxide is taken from the air by green plants, and put back into the air by all living organisms and by burning.

Photosynthesis uses up carbon dioxide

Only about 0.04 % of the air is carbon dioxide. Plants use carbon dioxide from the air when they photosynthesise. They use it to make food. The carbon from the carbon dioxide becomes part of the food molecules. Plants can only photosynthesise during daylight, because photosynthesis needs energy from sunlight.

Respiration produces carbon dioxide

All living things need energy. They get their energy from food. When the energy is released from food, carbon dioxide is produced. This is where the carbon dioxide which you breathe out comes from.

All livings things respire. So all living things produce carbon dioxide. Even plants produce carbon dioxide. They do it all the time. But during the day, they use it up in photosynthesis faster than they make it in respiration. So in the daytime plants take in carbon dioxide. At night they give it out.

Fig. 21.1 Carboniferous swamps. In the Carboniferous Period, between 280 and 320 million years ago, large areas of the Earth were covered with low-lying swamps, in which luxuriant vegetation grew in warm climates. As the plants died, they became squashed and compressed under the water, forming peat and then coal.

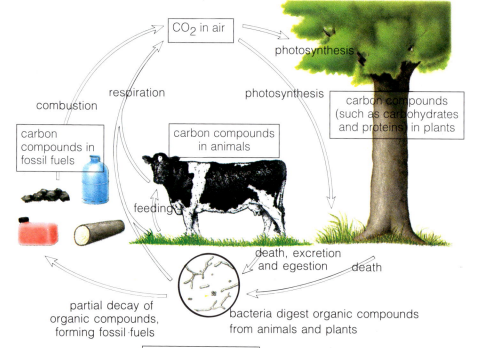

Fig. 21.2 The carbon cycle

Burning produces carbon dioxide

When things burn, they react with oxygen in the air. The fuels which we burn all contain carbon. The carbon reacts with oxygen in the air to form carbon dioxide.

How did the carbon get into the fuels? Fossil fuels, such as coal, oil and gas, were formed from plants and bacteria. The plants took carbon dioxide from the air. So the fuels contain carbon.

The enhanced greenhouse effect and global warming

Carbon dioxide in the air behaves rather like the glass of a greenhouse. It lets the Sun's rays through on to the surface of the Earth. Some of these rays heat the Earth's surface. Some of this heat escapes from the Earth and goes back into space. But the carbon dioxide in the atmosphere traps some of the heat and stops it escaping. This is called the **greenhouse effect**. Without the carbon dioxide, the Earth would lose much more heat and would be much colder than it is. Earth would be a frozen, lifeless planet.

Photosynthesis takes carbon dioxide out of the air. Respiration and burning put carbon dioxide into the air. These processes balance each other. So the amount of carbon dioxide in the air stays approximately the same.

But humans may now be upsetting the balance. We burn more and more fossil fuels. This releases extra carbon dioxide into the air, which increases the greenhouse effect. As a result, the Earth may be getting warmer. This is called **global warming**.

Does global warming matter? We do not know to what extent global warming will happen, or exactly what its effects will be. If it does happen, then it will change weather patterns on the Earth. Some places will become drier, while others will become wetter. Global warming may cause a lot of ice at the poles to become liquid water, which would increase sea levels. This could flood many major cities.

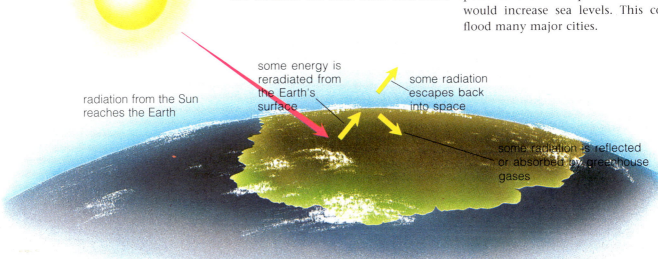

radiation from the Sun reaches the Earth

some energy is reradiated from the Earth's surface

some radiation escapes back into space

some radiation is reflected or absorbed by greenhouse gases

Fig. 21.3 The greenhouse effect. Radiation from the Sun passes through the atmosphere on to the Earth's surface. Here, some is reflected while some is absorbed and reradiated as heat. Some of this reradiated energy escapes into space, but some is retained in the atmosphere. Gases such as carbon dioxide and methane increase the amount of energy retained. This warms the Earth.

Fig. 21.4a Global surface air temperatures.
The horizontal dotted line shows the average air temperature on the Earth's surface. It is calculated from all the measurements taken, all over the world, between 1950 and 1979.
Eight of the nine warmest years this century have occurred in the 1980s. In 1988, the global average temperature was 0.34 °C above the long-term average. We do not know whether this has been caused by increased carbon dioxide emissions, or whether there is some other natural cause.

b Atmosphere carbon dioxide levels. In recent years the CO_2 level has begun to rise much more steeply than before. The levels of carbon dioxide in the 18th and 19th centuries have been measured from bubbles of air trapped in ice at the Poles.

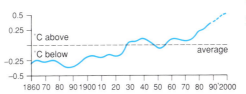

Questions

1 a What is the percentage of carbon dioxide in the atmosphere?

b Which process removes carbon dioxide from the atmosphere?

c Which processes return carbon dioxide to the atmosphere?

EXTENSION

2 Discuss the causes, and possible results, of global warming. What do you think that humans can do to keep the damage to a minimum?

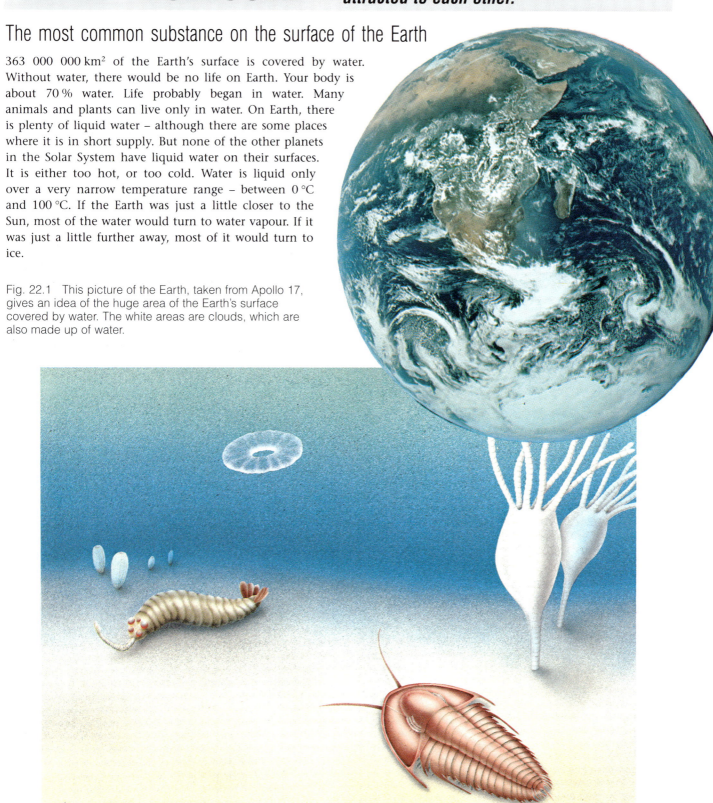

22 THE WATER MOLECULE

Water is a very strange substance. It has unusual properties because of the structure of its molecule. Water molecules are attracted to each other.

The most common substance on the surface of the Earth

363 000 000 km² of the Earth's surface is covered by water. Without water, there would be no life on Earth. Your body is about 70 % water. Life probably began in water. Many animals and plants can live only in water. On Earth, there is plenty of liquid water – although there are some places where it is in short supply. But none of the other planets in the Solar System have liquid water on their surfaces. It is either too hot, or too cold. Water is liquid only over a very narrow temperature range – between 0 °C and 100 °C. If the Earth was just a little closer to the Sun, most of the water would turn to water vapour. If it was just a little further away, most of it would turn to ice.

Fig. 22.1 This picture of the Earth, taken from Apollo 17, gives an idea of the huge area of the Earth's surface covered by water. The white areas are clouds, which are also made up of water.

Fig. 22.2 Life in the sea about 530 million years ago. At this time, all living things were aquatic. Nothing at all lived on land. Cells are mostly water, and die if they dry out. The first living organisms to have bodies which could retain water in their cells when on dry land, evolved about 400 million years ago.

Water molecules have dipoles

Water is a compound of hydrogen and oxygen. Each water molecule is made up of two hydrogen atoms, and one oxygen atom. The atoms are held strongly together by covalent bonds. The molecular formula for water is H_2O.

A covalent bond is an electron-sharing bond. Each hydrogen atom in a water molecule shares its electron with the oxygen atom. But the electrons are not spread evenly around the molecule. They spend more time close to the oxygen atom than the hydrogen atoms. Electrons have a negative charge. The oxygen atom in a water molecule gets more than its fair share of the electrons. So the oxygen atom has a slight negative charge. The hydrogen atoms get less than their fair share of the electrons. So they have a slight positive charge.

So one part of a water molecule has a slight negative charge, and two parts have a slight positive charge. This is called a **dipole**. Water molecules have dipoles.

It is important to remember that the *overall* charge on a water molecule is nil. The positive and negative charges are equal, so they cancel each other out.

Water molecules are attracted to one another

Figure 22.5 shows what happens in a group of water molecules. The positive parts of one water molecule are attracted to the negative parts of other water molecules. In an ice crystal, the water molecules arrange themselves in a regular pattern, or **lattice**. The attractions between the water molecules hold them in position.

If the ice is warmed, the water molecules begin vibrating faster. At 0 °C, they are moving vigorously enough to break out of the lattice pattern. They move freely around each other. The ice has melted. The water is now liquid. But the water molecules are still attracted to each other. As they move around, they are briefly attracted to each other. Liquid water is a mass of moving water molecules, briefly attracted to each other as they pass by.

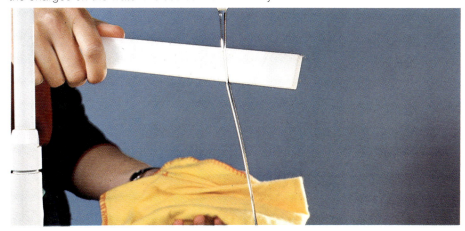

slight negative charge

slight positive charge

slight positive charge

Fig. 22.3 A water molecule has no overall charge. But there is a slight negative charge in the region of the oxygen atom, and a slight positive charge near the hydrogen atoms.

Fig. 22.4 This strip has been charged with static electricity by rubbing it with a cloth. The charge on the strip attracts the stream of water from the tap, because the charges on the water molecules are not evenly distributed.

Fig. 22.5

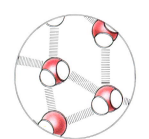

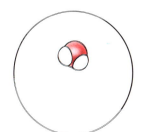

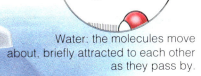

Ice: the water molecules are held firmly together in a lattice. The positive parts of each water molecule are attracted to the negative parts of its neighbours.

Steam: the water molecules are so far apart from each other that the forces between them are very small.

Water: the molecules move about, briefly attracted to each other as they pass by.

If the water is heated, the molecules move faster and faster. At 100 °C, they are moving so fast and so energetically that they can break away from each other completely. The attractive forces are no longer strong enough to hold them together. The water molecules fly away from one another, into the air. The water boils. It turns into gaseous water, or **steam.**

It is the attractive forces between water molecules which make it a liquid at normal temperatures. Most compounds with molecules similar to water, such as carbon dioxide and methane, are gases. Without the attractive forces between its molecules, water would be a gas at normal temperatures, and there would be no life on Earth.

Liquids often behave as if they have a weak 'skin'. This is caused by the attraction of the molecules in the liquid for each other.

Surface tension pulls water molecules together

A pond skater can walk on water. A dry needle can be made to float on a water surface. Why is this?

Water molecules are attracted to one another. In the middle of a container of liquid, a molecule has other molecules all around it. It is attracted equally from all directions. The forces balance out. At the surface, a molecule has no water molecules above it. So all the attractive forces are sideways and downwards. There is an overall force pulling the surface molecules inwards. This force keeps the surface together. The surface acts like a skin, with the molecules sticking together. This 'skin' is under tension, and the effect is known as **surface tension**.

It is surface tension which causes water to form droplets. In a raindrop, the molecules in the surface are pulled together. The tension in the surface pulls the surface into as small an area as possible. The shape with the smallest surface area for a given volume is a sphere. So water has a tendency to form spherical droplets. In the same way, liquid mercury forms small balls.

Fig. 23.2 Mosquito larvae breathing. Their water-repellent hairs break through the surface film.

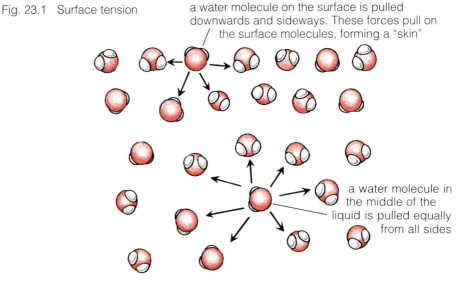

Fig. 23.1 Surface tension

a water molecule on the surface is pulled downwards and sideways. These forces pull on the surface molecules, forming a "skin"

a water molecule in the middle of the liquid is pulled equally from all sides

Many water animals rely on surface tension

Fig. 23.3 A pond skater, standing on the surface film of pond water.

Mosquito larvae live in water. But, like all insects, they have to get their oxygen from air. When they need air, they come to the surface 'tail first'. They float just under the water surface. Water-repellent hairs break through the surface film. The hairs are pulled apart by surface tension forces. Oxygen from the air can then diffuse into the larva.

If detergent is added to the water, it reduces the surface tension. The mosquito larvae's breathing method no longer works. Detergent is sprayed on to ponds and lakes in places where malaria is common. The detergent prevents the larvae from breathing. Mosquitos carry the organism which causes malaria, so killing mosquito larvae can help to keep malaria under control.

Other insects carry air stores under water with them. The water beetle *Dytiscus* does this. It holds a bubble of air underneath its wing cases. The surface tension of the water stops it from getting in between the water-repellent surfaces of the wing cases.

The camphor beetle lives on the surface film of small ponds. It walks on water. It produces a chemical from the tip of its abdomen which lowers the surface tension of the water behind it. The high surface tension at its head end now pulls it forward so it skims over the water like a tiny speedboat.

Surface tension prevents water from penetrating the tightly woven fibres of a tent canvas. Once the fragile film is broken by touching the canvas, the water drips through.

Attraction is not always between the same kind of molecules

Surface tension exists because molecules in a liquid are attracted to one another. This attraction is called **cohesion**. Cohesion makes water form droplets.

But water molecules are not only attracted to other water molecules. They may be attracted to other molecules. Molecules to which water molecules are attracted are called **hydrophilic** molecules. Glass is hydrophilic. Drops of rain cling to a glass window. The water molecules are attracted to the glass. This attraction is called **adhesion**. Water molecules are attracted to

Fig. 23.4   A drop of water spreads out on a clean sheet of glass, whereas mercury forms a rounded globule.

A meniscus is caused by adhesion and cohesion

Water in a glass container forms a **meniscus**. The shape of the meniscus is caused by adhesion between the water and the glass. The water is more strongly attracted to the glass than to itself. So where the water touches the

glass more than they are attracted to other water molecules. The adhesion is stronger than the cohesion. So the water wets the glass. It does not stay in round droplets, but spreads out over the surface of the glass.

Mercury behaves rather differently. Mercury molecules have stronger cohesion than adhesion to glass. Mercury particles are attracted more strongly to other mercury particles than they are attracted to glass particles. So mercury forms small droplets, and does not spread out on glass.

Fig. 23.5   Mercury forms a convex meniscus in a glass container, because its molecules are more attracted to each other than to the glass. Water forms a concave meniscus, because the molecules are more attracted to glass than to each other.

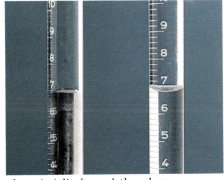

glass, it 'climbs up' the glass.

Mercury does the opposite. Cohesion between mercury particles is stronger than adhesion between mercury and the glass. So a mercury meniscus is 'upside down'.

Water creeps upwards by capillarity

If water is put into a narrow glass tube, adhesion between the water and the glass causes the water to rise up the edge of the glass. If the tube is narrow enough, the water level actually rises up the tube. The narrower the tube the higher the level it rises to. This is called **capillarity**.

Capillary action makes water creep between the fibres of a kitchen towel. Without capillary action, you could not mop up spills.

Soil particles are separated from each other by small air spaces. Soil particles are hydrophilic. They attract water molecules. So adhesion between water and soil particles is strong. Water moves up between the soil particles by capillarity. This is very important in providing plant roots with water. In soils where the spaces between the soil particles are too big, the water cannot rise very high. Plants may find it difficult to get enough water.

Capillarity can be a problem. Bricks contain tiny air spaces. Water can creep up through these spaces from the ground. This causes dampness in the house. A barrier called a **dampcourse** is put in just above ground level to stop this happening. It is made of a substance through which water cannot rise, such as tarred felt, slate or plastic.

Questions

1   Explain why:
   a Water forms a concave meniscus, while mercury forms a convex meniscus when in a glass container.
   b When water is poured out of a glass container the glass stays wet.
   c If a fly falls into water, it has great difficulty in getting out.
   d Water on a dirty car spreads out, but if the car is polished the water collects in beads.
   e Clay soils, made up of small particles, hold water better than sandy soils, which are made of large particles.

2   Could you mop up a spill of mercury with a cloth? Explain your answer.

3   The diagram shows a toy boat. It floats on water. A small pellet of soap is fixed at the back of the boat, so that it dips into the water. Soap reduces the surface tension of the water.

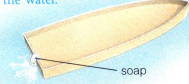
soap

   a Where is the surface tension greatest?
   b Are the forces on the boat balanced?

   c In which direction will the boat move?

4   A wire frame is dipped into soap solution. The lower film is then burst. What shape will the remaining film take up? Explain your answer in terms of surface tension.

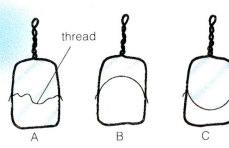
thread

A   B   C

A suspension is cloudy, and can be separated by filtration

If you stir some chalk into a beaker of water, you get a murky liquid. The tiny chalk particles float in the water. They may stay there for several hours, but most of them will eventually fall to the bottom of the beaker. Some, though, will stay floating in the water. They make it slightly cloudy. The mixture of chalk particles and water is called a **suspension**. Although the chalk particles may be too small for you to see, they are big enough to be trapped in filter paper. If you **filter** the suspension, you can separate the water and the chalk particles. The water is called the **filtrate**. The chalk particles on the filter paper are called the **residue**.

Fig. 24.1 A suspension of powdered chalk in water. Suspensions look cloudy.

Fig. 24.2 Separating powdered chalk from water by filtration. The clear liquid which runs through is the filtrate. The material which is trapped on the filter paper is the residue.

Fig. 24.3 Salt (sodium chloride) solution and copper sulphate solution. Solutions look clear.

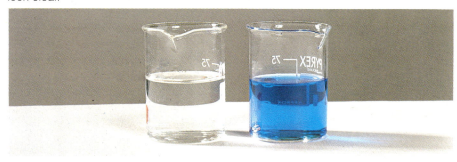

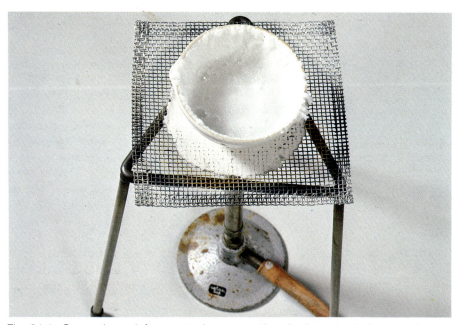
Fig. 24.4 Separating salt from water by evaporation. As the water boils away, the salt is left behind in the evaporating dish.

A solution is clear, and cannot be separated by filtration

If you add salt to a beaker of water, and stir, the salt seems to disappear. It forms a **solution**. The solution is clear. Some solutions, such as copper sulphate solution, are coloured. But they are always clear.

You cannot separate a salt solution by filtering. If you pour it through filter paper, the salt and the water both go through the paper. (How can you separate the salt and the water?)

In a solution, the substance which dissolves is called the **solute**. In salt water, salt is the solute. The liquid that the solute dissolves in is called the **solvent**. In salt water, water is the solvent.

Substances such as salt and copper sulphate, which will dissolve in water, are said to be **soluble** in water. Substances such as chalk particles are **insoluble** in water.

Water is an excellent solvent

A very large number of different substances will dissolve in water. Water is a very good solvent. Why is this?

To be able to answer this question, we need to understand what happens when something dissolves in water. Sodium chloride is a good example.

Sodium chloride, NACl, is an ionic compound. It is made up of sodium ions, Na⁺, and chloride ions, Cl⁻. The ions are held strongly together, because the positive and negative charges attract one another.

Water molecules have dipoles. Part of each water molecule has a slight negative charge, and part has a slight positive charge. When sodium chloride is stirred into water, the water molecules are attracted to the sodium ions and chloride ions. Figure 24.5 shows what happens. Each ion becomes surrounded by water molecules, attracted towards the ion. This is why the sodium chloride seems to disappear when it dissolves in water. Each sodium and chloride ion is separated from the others. They are in between the water molecules.

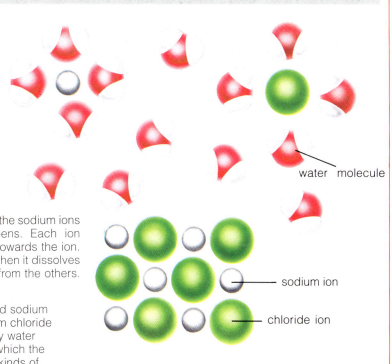

Fig. 24.5 Sodium chloride dissolving in water. In solid sodium chloride, each ion is held firmly in the lattice. If sodium chloride is mixed with water, each ion becomes surrounded by water molecules. Can you see the difference in the way in which the water molecules arrange themselves around the two kinds of ion? Can you explain this?

Immiscible liquids can form emulsions

When you add another liquid to water, it may mix with it completely. Ethanol will do this. Ethanol and water are said to be **miscible**. The ethanol dissolves in the water. A completely clear ethanol solution is formed.

Fig. 24.7 Separating petrol and water using a separating funnel. The water has been coloured so that it shows up more clearly.

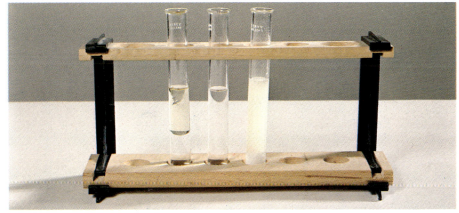

Fig. 24.6 Some liquids will mix together, while others will not. On the left, oil is floating on water – they do not mix. The centre tube contains a solution of methanol and water. Methanol is soluble in water. The tube on the right contains an emulsion of oil and water. The two liquids have been shaken strongly together, so that the oil has broken into tiny droplets which float in the water.

But if you add petrol to water, the two liquids will not mix. They are said to be **immiscible**. The density of petrol is lower than that of water, so the petrol floats on the water. Figure 24.7 shows how the petrol and water can be separated.

If you add cooking oil to water, the oil floats on the water. Oil and water are immiscible. But if you shake the oil and water mixture vigorously, you will end up with a milky-looking liquid. This is called an **emulsion**. As you shake the mixture of oil and water, you break up the oil into tiny droplets. The droplets are small enough to float in the water. They make the water look cloudy. Your emulsion of oil and water will probably separate if you leave it to stand. But many emulsions will stay for a long time. Milk is an emulsion. It is made up of tiny droplets of fat floating in water. Emulsion paint is another example. In this case, the tiny droplets in the water are coloured oils.

25 MORE ABOUT SOLUTIONS

Substances are more likely to react together when they are in solution than when they are in solid form.

Substances are more ready to react in solution than in solid form

If you mix dry copper sulphate crystals with dry sodium hydroxide, nothing happens. But when you mix a copper sulphate *solution* with a sodium hydroxide *solution*, they react together to form copper hydroxide and sodium sulphate. Why is this?

In solid copper sulphate, each Cu^{2+} ion is strongly attracted to the SO_4^{2-} ions around it. The bonds between the copper and sulphate ions are very strong. The ions are reluctant to move from their positions in the lattice.

The same is true of solid sodium hydroxide. The Na^+ ions are strongly attracted to the OH^- ions. So solid copper sulphate and solid sodium hydroxide do not react together.

But when they are in solution it is a different story. Now the ions are all separated from each other. Each ion is surrounded by a group of water molecules. Everything is continually moving around. So copper ions, sulphate ions, sodium ions, hydroxide ions and water molecules will keep bumping into each other.

When a copper ion bumps into a hydroxide ion, they are strongly attracted together. They form an insoluble compound, copper hydroxide. Because it is not soluble, the copper hydroxide makes the water look cloudy. The appearance of this pale blue cloudiness tells you that a reaction has happened in the solution. The copper hydroxide eventually falls to the bottom of the vessel and so is said to form a **precipitate**.

EXTENSION

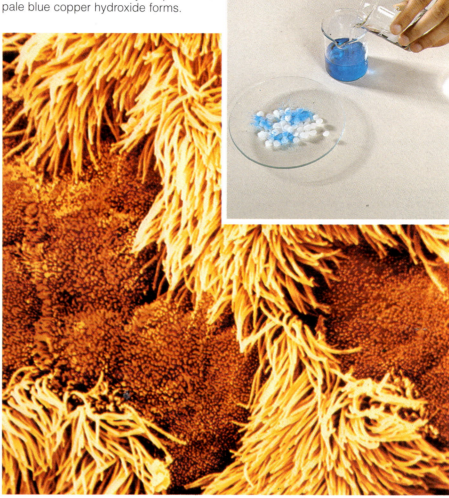

Fig. 25.1 Copper sulphate powder and sodium hydroxide pellets do not react together when dry. But if sodium hydroxide solution is added to copper sulphate solution, a cloudy suspension of pale blue copper hydroxide forms.

Fig. 25.2 Cells from the lining of the trachea. Like all cells, they are largely composed of water.

Reactions in living cells happen in solution

Your body is about 70 % water. If you lose much of this water, you suffer from dehydration. If a very large amount of water is lost, you may die.

One of the reasons you need so much water is to allow reactions to take place in your body. You are alive because chemical reactions are happening in your cells. There are hundreds of different reactions which are needed to keep you alive. One that you will meet in this book is **respiration**. Respiration is a chemical reaction which releases energy from food. It happens in every cell in your body.

These chemical reactions will happen only if the substances taking part in the reaction – the **reactants** – are in solution. Your cells are jelly-like solutions of many different substances in water. If the water was not there, the chemical reactions would not take place. You would quickly die.

DID YOU KNOW?

The longest time that anyone has ever lived without food and water is 18 days. An 18-year old man in Austria was put into a police cell on 1st April 1979, and forgotten. When he was discovered on 18th April, he was very close to death

Solubility is a measure of how much of a substance will dissolve

If you put 100 g of water into a beaker, add some sodium chloride and stir, it will dissolve. But if you go on and on adding sodium chloride, you will eventually get to a point where no more will dissolve. You have made a **saturated solution** of sodium chloride.

The amount of sodium chloride that will dissolve in 100 g of water is called the **solubility** of sodium chloride. Different substances have different solubilities.

Fig. 25.3 A SodaStream forces pressurised carbon dioxide into your drink. The colder the drink, the more carbon dioxide will dissolve, and the fizzier it will be.

Solubility of solids increases with increasing temperature; solubility of gases decreases with increasing temperature

If you heat a saturated solution of sodium chloride, you will find that you can 'persuade' a little more sodium chloride to dissolve in it. The solubility of sodium chloride increases as the temperature increases. This is true for most solids. The amount of a substance which can dissolve in 100 g of water at different temperatures can be shown on a graph called a **solubility curve.**

Gases, however, become *less* soluble at higher temperatures. Oxygen is slightly soluble in water. Fish and other water animals use this dissolved oxygen. They take it in through their gills, and use it for respiration. In a pond on a cool day, there is usually plenty of oxygen dissolved in the water. But on a hot day, the hot, fast-moving oxygen molecules can escape from the water into the air. Less oxygen remains in solution in the water. Fish may get short of oxygen, and have to come to the surface to get air.

Fig. 25.4 Waste water from a coal- and oil-burning power station discharges into the River Medway. The water is warm, creating problems for living organisms in the river.

Questions

1 Explain what is meant by the terms:
   a precipitate **b** saturated solution
   c solubility **d** solubility curve
2 Describe, in your own words, why substances are more likely to react together when they are in solution, than when they are in solid form.
3 The chart shows the solubility of copper (II) sulphate at different temperatures.

a Draw a solubility curve for copper (II) sulphate. Temperature goes on the horizontal axis.

b What would happen if you cooled a saturated solution of copper (II) sulphate from 70 °C to 10 °C?

Temperature (°C)	0	10	20	30	40	50	60	70
Number of grams which will dissolve in 100 g of water	14	17	20	25	29	34	40	47

Soaps and other detergents are used to help water to remove dirt. They work by reducing surface tension.

Soap was the first detergent

When water is used for washing, it dissolves dirt. But some dirt will not dissolve in water. Oily dirt will not dissolve. Detergents help oil to dissolve in water. Soap was the first detergent. People have been using soap for a very long time. The first kinds of soap were made by boiling animal fat with wood ash. This is how the Romans made soap. Now most soap is made from oils which smell more pleasant, such as palm oil. The oil is boiled with potassium hydroxide solution.

Fig. 26.1 Soaps, made from plant and animal fats.

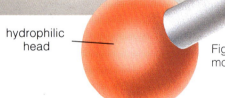

INVESTIGATION 26.1

How does detergent affect surface tension?

1 Take a small piece of fabric. Using a dropper pipette, carefully place one drop of water on to its surface. Draw a diagram of the water droplet on the cloth.
2 Take a second piece of the same fabric. Mix a little detergent with water in a beaker. Carefully place one drop of the detergent solution on to the surface of the cloth. Draw a diagram of the water on the cloth.

Questions

1 Why does the water without detergent form a rounded droplet on the cloth?
2 What happens to the drop of water plus detergent on the cloth?
3 Use your answers to questions **1** and **2** to explain how detergents can help when washing clothes.

Detergent molecules are attracted to both water and oil

Oil will not dissolve in water. Oil and water molecules are not attracted to one another. Oil molecules are said to be **hydrophobic**. Hydrophobic means 'water-hating'.

Detergent molecules are made up of two parts. They have a head and a tail. The tail is hydrophobic. It is not attracted to water molecules. But it *is* attracted to oil. The head of a detergent molecule is attracted to water molecules. The head is **hydrophilic**. Hydrophilic means 'water-loving'.

Figure 26.2 shows how detergent molecules can help oil to float away in water. The hydrophobic tails are attracted to the oil, and bury themselves in it. The hydrophilic heads are attracted to the water. The detergent molecules pull the oil into a rounded droplet which can float in the water. An emulsion is formed. An emulsion is a mixture of tiny droplets of one liquid in another. When you wash a greasy plate with a solution of washing-up liquid, this is what happens. The detergent helps the grease to form an emulsion in your washing-up water. The grease comes off the plate, and into the water.

hydrophobic tail

hydrophilic head

Fig. 26.2a A detergent molecule

detergent molecule

water

oil drop

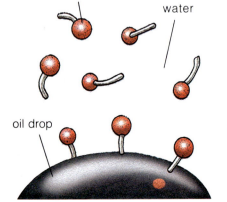

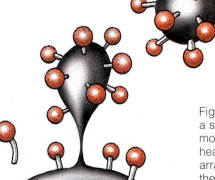

Fig. 26.2b Detergent removing grease (oil) from a surface. The hydrophobic tails of the detergent molecules are attracted to oil. The hydrophilic heads are attracted to water, so the molecules arrange themselves with their tails in the oil and their heads in the water. The heads repel each other, swinging apart and pulling the oil into a spherical droplet. The droplet of oil, surrounded by detergent molecules, floats off into the water.

Fig. 26.3 Many foods contain emulsifiers. These stop fat and water in the food from separating out into layers.

Fig. 26.4 Most products which we use for washing are soapless detergents.

Bile contains detergents

A similar thing happens in your intestine. Fat or oil in your food will not dissolve in the watery liquids inside your digestive system. So a liquid called **bile** is poured into your intestine. Bile contains detergent-like substances called **bile salts**. Bile salts help the fat in your food to form tiny droplets in the liquids inside your intestine. They are said to **emulsify** the fats. This makes it much easier for the fat to be digested.

Many modern detergents are soapless detergents

Early detergents were made from sodium or potassium hydroxide and animal or plant oils. But animal and plant oils are better used for human food. Now, most detergents are made using by-products of oil refining. They are called **soapless detergents**. Soapless detergents are better than soap in another way, too. Soap forms scum with hard water. This wastes a lot of soap. Soapless detergents do not form a scum.

Fig. 26.5 Foaming caused by detergent pollution.

Detergents and pollution

The first soapless detergents caused pollution problems. They could not be broken down by bacteria. They left houses and factories in waste water, and flowed into the sewage system. Sewage treatment did not break them down. So they eventually got into rivers and streams. Here, they caused foaming, which looks ugly and can harm animals and plants in the water.

Modern soapless detergents can be broken down by bacteria. They are called **biodegradable** detergents. The bacteria in sewage treatment plants break them down. So modern detergents are less likely to cause foaming in rivers.

But detergents do still cause pollution problems. Many detergents have phosphate salts added to them. This improves their dirt-removing power. These phosphates can get into streams and rivers. They act as fertilisers, so that water plants grow more than usual. When the plants die, bacteria in the rivers feed on them. The bacteria use up all the oxygen in the water, so that there is none for fish and other animals.

Questions

1 a What is a detergent?
b Explain the difference between soap and soapless detergents.
c What are the advantages of soapless detergents?
2 a What is an emulsion?
b How do bile salts help you to digest fatty foods?
c Emulsifiers are often added to foods which are made of mixtures of watery and fatty substances. Ice cream is an example. What do these emulsifiers do? What might happen if they were not added?
3 Explain how detergents can cause pollution. What can be done to reduce this problem?

Questions

1 Read the following information, and then answer the questions. Make your answers to each question as full as possible.

THE HOLE IN THE OZONE LAYER

Ozone is a gas. Its molecules are made of three oxygen atoms linked together, so its molecular formula is O_3. Ozone is formed from normal oxygen, O_2, in ultra-violet light. It can also be broken down in ultra-violet light. In the upper atmosphere, the creation and breaking down of ozone balance each other out, so the amount of ozone tends to stay about the same.

A map drawn from information collected by the American Nimbus-7 weather satellite in October, 1987. The scale shows the amount of ozone in the atmosphere, measured in Dobson units. The pink and dark blue areas over Antarctica contain very little ozone.

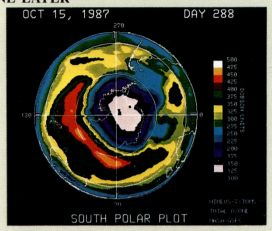

Ozone is produced at ground level, too. You may have noticed the smell of ozone when a toy car sparks, or when you use some types of battery charger, or a photocopier. Ozone is also produced when sunlight shines on car exhaust fumes.

The layer of ozone in the upper atmosphere is very useful to us. It stops too much ultra-violet light from reaching the Earth's surface. Ultra-violet light can damage cells, causing mutations and possibly skin cancer.

But we are damaging the protective ozone layer. Substances called chlorofluorocarbons, or CFCs, are used in many aerosols. They are also used in making fast-food cartons and cavity-wall insulation, in refrigeration systems, and in air conditioning. CFCs diffuse up into the upper layers of the atmosphere. Here, ultra-violet light acts on them to produce chlorine. The chlorine reacts with ozone, breaking it down. There is now a hole in the ozone layer. It appears over the Antarctic, and is largest in the spring each year. This is very worrying. If the amounts of ozone keep going down, we will be at danger from ultra-violet light.

People are now trying hard to stop the situation getting any worse. The amount of CFCs used is being reduced. In October 1987, an international agreement was signed in Canada to reduce CFC emissions by 50% of the 1986 levels by 1999. But not every country signed this agreement. And even if there was a complete ban beginning today, the problem would still not go away. CFCs may persist in the atmosphere for centuries.

a What is ozone?
b Where is the ozone layer?
c What use is the ozone layer?
d Why has a hole appeared in the ozone layer?
e What is being done to stop the problem getting any worse?

2 Explain each of the following:
a When liquid air is warmed, nitrogen turns to a gas before oxygen does.
b The concentration of carbon dioxide in the air above a grass field goes down during the day.
c Nitrogen is often used for food packaging.
d Carbon dioxide is used in fire extinguishers.

CHEMICAL REACTIONS

27 ACIDS AND ALKALIS

Acids are sour-tasting liquids. Alkalis can neutralise acids. The pH of a liquid is a measure of how acidic or alkaline it is.

Acids are sour-tasting liquids

Acids taste sour. Some common examples include lemon juice and vinegar. These are **weak** acids. But some of the acids that you use in the laboratory are **strong** acids. These include hydrochloric acid and sulphuric acid. Acids are corrosive. Strong acids can rapidly burn your skin.

Tasting a liquid is not a good way to find out if it is an acid! A much better way is to use an **indicator**. An indicator is a substance which changes colour depending on whether a substance is an acid or an alkali. Many natural substances are good indicators. Red cabbage juice and blackcurrant juice are two examples. **Litmus** is a purple dye which comes from lichens. It turns red in acids. Litmus paper is paper which has been soaked in litmus solution and then dried.

Fig. 27.5 Red cabbage juice is a natural indicator. The beaker contains a liquid obtained by liquidising red cabbage leaves. The juice has been pipetted into an alkali on the left, tap water (neutral) in the centre, and acid on the right.

Fig. 27.1 Some examples of weak acids

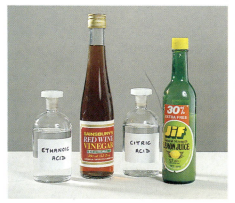

Fig. 27.2 Strong acids

Fig. 27.3 Some examples of weak alkalis

Fig. 27.4 Strong alkalis

Alkalis turn litmus paper blue

Alkalis are the chemical 'opposite' to acids. Some examples of alkalis are potassium hydroxide, calcium hydroxide, sodium hydroxide and ammonium hydroxide. Like acids, strong alkalis can burn your skin. They feel 'soapy' if you get them on your skin. This is because the alkali reacts with oils in your skin to make soap. You can test for alkalis using litmus paper. It turns blue in an alkaline solution.

Alkalis are soluble in water. A more general name for a substance that is 'opposite' to an acid is **base**. An alkali is therefore a soluble base. Most alkalis are the hydroxide compounds of metals. Ammonium hydroxide is an exception to this.

70

The strength of an acid or alkali is measured on the pH scale

The **pH scale** runs from 0 to 14. A substance with a pH of 7 is **neutral**. This means that it is neither acidic nor alkaline. Water is the best-known example of a neutral substance. It is neither acidic nor alkaline and therefore has a pH of 7. (And by the way, 'pH' isn't a misprint, it really is meant to be a small p and a capital H.) A substance with a pH of below 7 is acidic. If it has a pH of above 7, it is alkaline. Strongly acidic substances have a very low pH. The lower the pH, the stronger the acid. Similarly, the higher the pH, the stronger the alkali.

pH can be measured using a pH meter. This is simply dipped into the solution you are testing, and gives a reading on a digital display. But pH meters are expensive and can be delicate. A more common method of measuring pH in a school laboratory is to use **universal indicator**. This is a mixture of indicators, which has different colours across the entire pH range. You can use it as a liquid or as paper.

Fig. 27.6 A pH meter. The electrode is placed into the solution whose pH you want to measure.

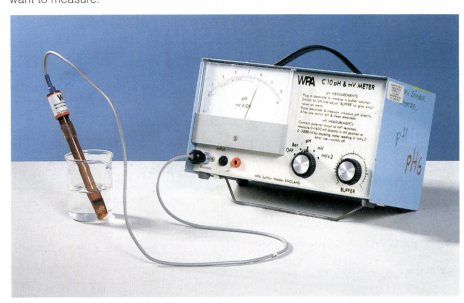

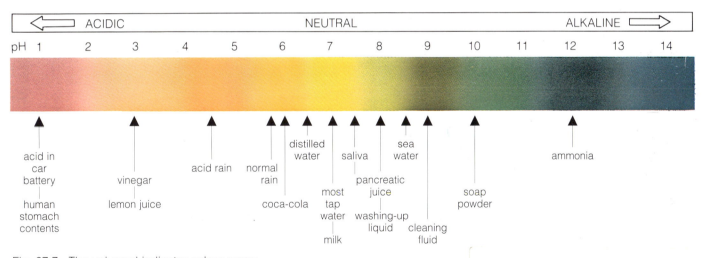

Fig. 27.7 The universal indicator colour range

Questions

1 Divide part of a page in your book into two columns. Head one column ACIDS and the other column ALKALIS. Then decide which of the following pairs of statements fits in each column, and write them in. Keep the two statements in a pair opposite each other in your chart.

a **A** taste sour
  B feel soapy on your skin

b **A** turn litmus paper blue
  B turn litmus paper red

c **A** turn universal indicator green to purple
  B turn universal indicator orange to red

d **A** have a pH of over 7
  B have a pH of below 7

Acids are acidic because they contain H⁺ ions

All acids contain hydrogen ions (H^+). This is what makes them acids. The more hydrogen ions they contain, the more acidic they are. A good definition of an acid is a *substance which produces H^+ ions when it dissolves in water.*

Strong acids produce a lot of H⁺ ions

Hydrogen chloride is a colourless gas. It has no effect on dry litmus paper. It is not acidic. But when it dissolves in water, it forms **hydrochloric acid**. Hydrochloric acid is a strong acid. It is very corrosive, and can burn your skin. It turns litmus paper and universal indicator solution red. It has a pH of 1–3.

What happens when hydrogen chloride dissolves in water? As the hydrogen chloride molecules dissolve, they split or **dissociate** into hydrogen ions and chloride ions. We can show this as follows:

$$HCl(aq) \rightarrow H^+(aq) + Cl^-(aq)$$

The (aq) means 'dissolved in water'. 'Aq' is short for 'aqueous', which means 'to do with water'. Hydrogen chloride dissociates very easily when it dissolves in water. It produces a lot of hydrogen ions. This means that it is a **strong** acid.

Weak acids produce only a few H⁺ ions

[...]on juice contains citric acid. Citric [...]olecules also dissociate when [...]olve in water. They also pro-[...]en ions. But they do not [...] easily. So they do not [...] hydrogen ions. Cit-[...]id. It has a pH of [...]ersal indicator

Both strong and weak acids can be concentrated or dilute

If a lot of hydrogen chloride dissolves in a small amount of water, a concentrated solution is made. It is called **concentrated hydrochloric acid**. If a small amount of hydrogen chloride dissolves in a lot of water, a dilute solution is made. It is called **dilute hydrochloric acid**. In the same way, you can have either concentrated or dilute solutions of a weak acid like citric acid.

The **strength** of an acid tells you how easily it dissociates to produce hydrogen ions. The **concentration** of an acid tells you how much water it contains. Both strong and weak acids can be dilute or concentrated. Concentrated acids can be made dilute by adding water. Dilute acids can be made concentrated by evaporating off most of the water in them.

Fig. 28.1 Strong, weak, concentrated and dilute acids. The water molecules are not shown.

STRONG ACID e.g. HCl(aq)

WEAK ACID e.g. CH₃COOH(aq)

CONCENTRATED - a lot of hydrogen ions and negative ions in a certain volume of water

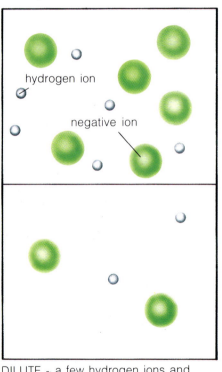
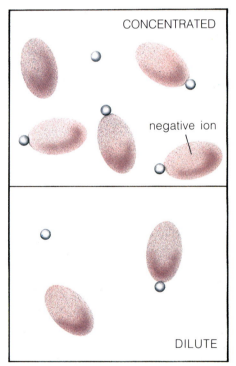

DILUTE - a few hydrogen ions and negative ions in a certain volume of water

Hydrogen chloride forms ions in solution

Hydrogen chloride gas is a covalent compound. The hydrogen atom and chlorine atom share electrons. They are held together by a covalent bond.

But when hydrogen chloride dissolves in water, the hydrogen and chlorine atoms separate. The hydrogen atoms lose one electron each, becoming hydrogen ions. The chlorine atoms gain one electron each, becoming chloride ions. So some covalent compounds can form ions when dissolved in water!

EXTENSION

Strong and weak acids

In this investigation you are going to compare two acids. Hydrochloric acid, HCl(aq), is a strong acid – it produces a high concentration of H^+ ions. Ethanoic acid, $CH_3COOH(aq)$, is a weak acid – it produces a lower concentration of H^+ ions.

1 Put a piece of magnesium ribbon into each of two test tubes. Cover one piece with hydrochloric acid, and the other with ethanoic acid. Compare the rates of reaction.

2 Repeat 1, but using marble instead of magnesium.

3 Pour HCl into a test tube to depth of 1 cm. Add one drop of universal indicator solution. Note the colour and pH.

4 Mix $1\,cm^3$ of HCl and $9\,cm^3$ of water in a test tube. This is Solution A. Test its pH as in 3.

5 Mix $1\,cm^3$ Solution A and $9\,cm^3$ water. This is Solution B. Test its pH.

6 Mix $1\,cm^3$ Solution B and $9\,cm^3$ water. This is Solution C. Test its pH.

7 Repeat steps 3–6 using ethanoic acid instead of hydrochloric acid.

8 Write up all your results in the way you think is best. You might like to use a comparison table.

9 Summarise the evidence you have for saying that ethanoic acid is a weaker acid than hydrochloric acid.

DID YOU KNOW?

Hydrofluoric acid is very corrosive. Concentrated hydrofluoric acid eats into glass. Although dangerously corrosive to skin, it has been used to etch glass, where it forms silicon tetrafluoride.

All alkalis contain OH⁻ ions

An alkali is a substance which produces hydroxide (OH^-) ions when it dissolves in water.

Like acids, alkalis can be strong or weak. The more hydroxide ions it produces, the stronger the alkali.

Sodium hydroxide is an example of a strong alkali. When it dissolves in water, it readily dissociates to produce a lot of hydroxide ions.

$$NaOH\,(aq) \rightarrow Na^+\,(aq) + OH^-(aq)$$

Strong alkalis have a pH of 12–14. They turn universal indicator dark blue or purple.

Ammonia solution is an example of a weak alkali. When ammonia dissolves in water, the ammonia molecules react with the water molecules to form ammonium ions and hydroxide ions:

$$NH_3(aq) + H_2O(l) \rightarrow NH_4^+(aq) + OH^-(aq)$$

However, only a small proportion of the ammonia molecules do this, so only a low concentration of OH^- ions are produced. So ammonia solution is a weak alkali. Weak alkalis normally have a pH of 8–11. They turn universal indicator dark green or blue. The (l) after water is a **state symbol**. It tells you that the water is a liquid. You have already met the state symbol (aq) for aqueous or dissolved in water. Other state symbols commonly used are (s) for solid and (g) for gas.

Measuring the pH of soil

The pH of soil affects the plants which grow in it. Farmers and gardeners need to know how acidic or alkaline their soil is, so that they know which plants they can grow.

You cannot just add universal indicator to a mixture of soil and water, because the colour of the soil muddles your result. You need to filter the soil first.

1 Take your first soil sample. Put two spatulas of it into a test tube. Add $10\,cm^3$ of water. Cork the tube, and shake really well to mix.

2 Set up a filter funnel and paper as shown in Figure 28.2. Filter your soil and water mixture.

3 Test the filtrate with universal indicator, and record the result.

4 Repeat with your other soil samples.

Questions

1 Does it matter what sort of water you add to your soil sample? Why?

2 Most plants which we grow as vegetables, such as cabbages and carrots, like a neutral soil. Is your local soil suitable for growing these vegetables?

3 Do any of your soil samples have an acid pH? If so, what sort of soil is it? Where does it come from?

Fig. 28.2 Filtering a soil suspension

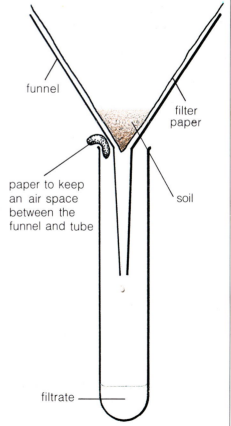

funnel

filter paper

paper to keep an air space between the funnel and tube

soil

filtrate

29

NEUTRALISATION OF ACIDS BY ALKALIS

Acids and alkalis readily react together. The alkali neutralises the acid, and a new substance called a salt is formed.

Alkalis neutralise acids

If you add hydrochloric acid to some sodium hydroxide solution in a beaker, the solution gets hot. If you use equal amounts of equal concentrations of acid and alkali, you will find that the resulting solution does not affect litmus paper. It has a pH of 7. It is a neutral solution.

On page 72, you saw that a solution of hydrochloric acid contains H^+ and Cl^- ions. Sodium hydroxide solution contains Na^+ and OH^- ions. When these four kinds of ions are mixed up together in the beaker, the H^+ and OH^- ions react together to make water molecules:-

$$H^+(aq) + OH^-(aq) \rightarrow H_2O(l)$$

If there were equal amounts of H^+ and OH^- ions in the acid and alkali you mixed up, then all the H^+ ions from the acid will react like this. The H^+ ions have been removed from the acid. The acid has been neutralised.

Equations show what happens during a chemical reaction

So far, we have only looked at what has happened to two of the four kinds of ions which were mixed in the beaker. What about the Na^+ and Cl^- ions?

They do not really do anything at all. They just remain as ions, dissolved in the water in the beaker. But, if you heated the solution in the beaker so that the water evaporated, the Na^+ and Cl^- ions would form an ionic lattice. You would have **sodium chloride** – common salt – in the bottom of your beaker. Any ionic compound formed in this way (by replacing hydrogen ions in an acid with other positive ions) is called a **salt**. Sodium chloride is one of many different salts.

We can write a full equation for this reaction, including the sodium and chloride ions as well as the hydrogen and hydroxide ions, like this:

$$H^+(aq) + Cl^-(aq) + Na^+(aq) + OH^-(aq) \rightarrow Na^+(aq) + Cl^-(aq) + H_2O(l)$$

This is called an **ionic equation**. It shows what is happening to each kind of ion in the reaction.

We can also show the same reaction in terms of the molecules which are taking part in it:

$$HCl(aq) + NaOH(aq) \rightarrow NaCl(aq) + H_2O(l)$$

This is called a **molecular equation**.

Yet another way of showing what happens in the reaction is in words, without using symbols or formulae at all, like this:

hydrochloric acid + sodium hydroxide $\rightarrow$ sodium chloride + water

This is called a word equation.

Word equations, molecular equations and ionic equations all tell you what happens in a reaction. But molecular equations or ionic equations give much more information than word equations.

INVESTIGATION 29.1

Neutralising a weak acid with a weak alkali

In this investigation you will mix a weak acid (ethanoic acid) with a weak alkali (ammonia solution), and follow the changes with universal indicator solution.

1 Read through the investigation, and decide how you are going to record your results. Draw up a results chart if you decide to use one.

2 Fill approximately one-quarter of a boiling tube with ethanoic acid.

3 Fill another boiling tube approximately two-thirds full with ammonia solution.

4 Add universal indicator solution to each, to get a distinct, but not-deep, colour.

5 Use a teat pipette to add the ammonia solution to the ethanoic acid drop by drop. After every ten drops, stir the mixture with a glass rod, and note the colour of the indicator and the pH number of the mixture.

Questions

1 Ethanoic acid is a weak acid, and ammonia solution is a weak alkali. What is meant by the word 'weak'?

2 What was the pH of the ethanoic acid?

3 What was the pH of the ammonia solution?

4 What happened to the pH of the acid solution as you added alkali to it?

5 Did you, at any stage, produce a neutral solution? If so, how many drops of ammonia solution did it take to produce this solution?

Neutralising a strong acid with a strong alkali

In this investigation you will mix a strong acid (hydrochloric acid) with a strong alkali (sodium hydroxide solution), and follow one of the changes with a thermometer.

1 Fill the burette to zero with the acid.
2 Use a measuring cylinder or pipette to put 20 cm³ NaOH(aq) into a conical flask.
3 Measure and write down the temperature of the NaOH(aq).
4 From the burette, add 2 cm³ of acid to the conical flask. Swirl the mixture. Measure and write down the new temperature of the alkali in the flask.

5 Continue adding acid to the flask, 2 cm³ at a time. Swirl the flask, measure and record the temperature after every addition. Do this until you have added a total of 40 cm³ of acid.
6 Thoroughly wash out the apparatus.
7 Draw a line graph of your results. Put volume of acid added on the horizontal axis and temperature on the vertical axis.

You will have seen that as acid is added to the alkali heat is given out. As the acid and alkali react, they release heat energy. A reaction like this one, which gives out heat, is said to be an exothermic reaction.

Questions

1 What evidence have you that a chemical reaction took place as the acid and alkali were mixed? Your results should show that the temperature gradually rose, and then gradually fell.
2 What was the highest temperature reached?
3 How much acid had been added in total at this point?
4 Explain why the temperature fell after this point.

Questions

Read the following information about tooth decay, and then answer the questions.

1 Explain what is meant by:
a dental caries
b plaque
c abscess
   The graph shows how the pH of someone's mouth changed during one day.

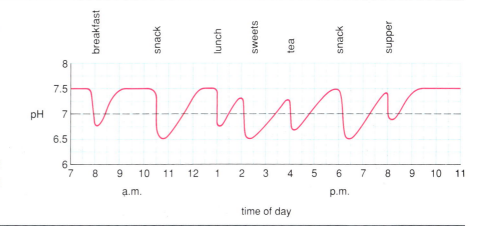

time of day

Tooth decay

Bad teeth can make you miserable. They look awful, can cause bad breath – and they hurt! Tooth decay, or dental caries, is caused by bacteria. Everyone has bacteria living in their mouth. If sugary foods are left on, or between, your teeth, some of these bacteria feed on it. They form a sticky film on your teeth, called plaque. The bacteria change the sugar into acid. The acid eats through the enamel covering of the teeth. If the amount of damage is slight, the tooth can repair itself. But if a tooth is exposed to acid for a long time, then a hole gradually develops in the enamel. Once it reaches the dentine, the hole enlarges rapidly. It may begin to be painful. If the decay reaches the pulp cavity, it is very painful. An abscess may result, which is an extremely painful, infected swelling beneath the tooth.

None of this need happen to you! If you never ate sweet foods, you would never get tooth decay. But most people like sweet things sometimes. And it is possible to eat them, and still enjoy perfect teeth. One of the golden rules is to eat your favourite sweet foods – and drink sugar-containing drinks – only two or three times a day. If your teeth are exposed to acid all day long, they do not have a hope of fighting off decay. But if the conditions in your mouth are acidic for only a short while, decay is less likely to happen. Another thing that you can do is to brush the plaque off your teeth twice a day. If you use a fluoride toothpaste, the fluoride will help your teeth to resist decay. Toothpastes also contain weak alkalis, which will neutralise the acid formed by bacteria in your mouth.

2 When a solution becomes more acidic, does the pH get higher or lower?
3 Explain why the pH in the person's mouth became lower after each meal.
4 What damage could be caused while the pH is low?
5 How could the person change their eating habits to lessen their chances of suffering from tooth decay?
6 Give three reasons why brushing regularly with fluoride toothpaste can lower the likelihood of suffering from tooth decay.

30

BALANCING EQUATIONS *Molecular equations show the formulae of the substances involved in a chemical reaction. The numbers of each kind of atom must balance on each side of the equation.*

Equations must be balanced

You have already seen that hydrochloric acid and sodium hydroxide solution react together to form sodium chloride and water. The word equation for this reaction is:

hydrochloric acid + sodium hydroxide → sodium chloride + water

The molecular equation for this reaction is:

$$HCl(aq) + NaOH(aq) \rightarrow NaCl(aq) + H_2O(l)$$

The molecular equation tells us more than the word equation. It tells us **how many** of each kind of atom react

Fig. 30.1 Sodium hydroxide and hydrochloric acid react to form sodium chloride and water.

together. Look first at the number of hydrogen atoms. On the left-hand side of the equation, there is one hydrogen atom in the HCl molecule, and one in the NaOH molecule. On the right-hand side of the equation, there are also two hydrogen atoms. They are both in the water molecule.

Check the numbers of each of the other kinds of atoms – chlorine, oxygen, and sodium – on each side of the equation. You should find that, for each of them, the numbers are the same on the left and the right. A molecular equation in which the numbers of each kind of atom are the same on the left and the right hand side is called a **balanced equation**. Equations must always be balanced.

NaOH + HCl ⟶ NaCl + H₂O

You sometimes need to put numbers in front of formulae to make equations balance

Another example of the neutralisation of an acid by an alkali is the reaction between **sodium hydroxide solution** and **sulphuric acid**.

The word equation for this reaction is:

sulphuric acid + sodium hydroxide → sodium sulphate + water

If we now begin to write a molecular equation, by writing chemical formulae instead of words, we get:

$$H_2SO_4 + NaOH \rightarrow Na_2SO_4 + H_2O$$

But there is something badly wrong with this equation. Look at the numbers of hydrogen atoms first of all. On the **left** side of the equation, there are two hydrogen atoms in the H_2SO_4 molecule, and one hydrogen atom in the NaOH molecule. That makes **three**

altogether. But on the right side of the equation, there are only **two** hydrogen atoms, both in the H_2O molecule. So this cannot be right! Hydrogen atoms do not just disappear like that. You cannot start off with three and end up with two! And if you look again at the equation, you will see that the numbers of sodium atoms do not balance out either. There are two on the right-hand side, but only one on the left-hand side. An extra sodium atom seems to have appeared from nowhere – which just is not possible. So something must be done to the equation to make it **balance** – to make the numbers of each kind of atom match up on either side.

Let us get the sodium atoms right first. There are two on the right-hand side of the equation, as sodium sulphate has the formula Na_2SO_4. So we need two sodium atoms on the left-hand side as well. We can do this by having **two** molecules of NaOH taking

part in the reaction:

$$H_2SO_4 + 2NaOH \rightarrow Na_2SO_4 + H_2O$$

But now the hydrogens and oxygens are wrong. There are four hydrogens on the left, and only two on the right. There are six oxygens on the left, and only five on the right. (Can you find them all?) We can sort this out in one go by having **two** water molecules on the right-hand side of the equation:

$$H_2SO_4 + 2NaOH \rightarrow Na_2SO_4 + 2H_2O$$

And that is the correct, balanced equation. If you add up the numbers of each kind of atom on each side of the equation, you should find that they match. Check it for yourself.

Equations must always balance. You must not lose or gain atoms between the beginning and end of a chemical reaction. The number of a particular sort of atom on one side of an equation must always equal the number of that sort of atom on the other side.

Another example of balancing an equation

When nitric acid reacts with calcium hydroxide solution, calcium nitrate and water are formed. How do you go about writing a balanced equation for this reaction?

First, write a word equation:

nitric acid + calcium hydroxide → calcium nitrate + water

Next, write down the correct formulae for each of the substances involved in the reaction. It is very important that you get these right. They are:

$$HNO_3 + Ca(OH)_2 \rightarrow Ca(NO_3)_2 + H_2O$$

Next, count up numbers of atoms on either side of the equation.

Calcium seems all right – there is one calcium atom on each side. But the hydrogen is *not* right. HNO_3 has one, and $Ca(OH)_2$ has two, so there are three hydrogen atoms on the left-hand side. On the right, there are only two hydrogen atoms. (Where are they?)

Count up the numbers of nitrogen and oxygen atoms. Do they balance?

The next thing to do is to *put a number in front of one of the formulae,* and see if it helps. Until you really get the hang of this, it is best just to guess and try it out. You can always rub it out if it does not work. In this case, it seems sensible to leave the $Ca(OH)_2$ and $Ca(NO_3)_2$ alone, as the Ca atoms balance already. So let us try:

$$2HNO_3 + Ca(OH)_2 \rightarrow Ca(NO_3)_2 + H_2O$$

Count atoms again. You should find that calcium and nitrogen balance, but that hydrogen and oxygen do not. On the left, there is a total of four hydrogen atoms, but only two on the right. The left has eight oxygen atoms, but only seven on the right. We can sort this out by putting a two in front of the water molecule:

$$2HNO_3 + Ca(OH)_2 \rightarrow Ca(NO_3)_2 + 2H_2O$$

Now the equation balances. Finish it off by putting in the state of each of the molecules involved:

$$2HNO_3(aq) + Ca(OH)_2(aq) \rightarrow Ca(NO_3)_2(aq) + 2H_2O(l)$$

Rules for balancing equations

1 Write a word equation.
sulphuric acid + sodium hydroxide →
sodium sulphate + water

2 Write correct molecular formulae.
$H_2SO_4 + NaOH \rightarrow Na_2SO_4 + H_2O$

3 Balance with numbers in front of formulae if necessary.
$H_2SO_4 + 2NaOH \rightarrow Na_2SO_4 + 2H_2O$

4 Add state symbols.
$H_2SO_4(aq) + 2NaOH(aq) \rightarrow Na_2SO_4(aq) + 2H_2O(l)$

Questions

1 Hydrochloric acid and potassium hydroxide react and produce potassium chloride and water.
a Write a word equation for this reaction.
b Write an equation with the formula of each substance.
c Add large numbers in front of the formulae if necessary to balance the equation.
d Add correct state symbols after the formulae.

2 Ammonium hydroxide and nitric acid react and produce ammonium nitrate and water.
a Write a word equation for this reaction.
b Write an equation with the formula of each substance.
c Add large numbers in front of the formulae if necessary to balance the equation.
d Add correct state symbols after the formulae.

3 Sulphuric acid and magnesium hydroxide react and produce magnesium sulphate and water.
a Write a word equation for this reaction.
b Write an equation with the formula of each substance.
c Add large numbers in front of the formulae if necessary to balance the equation.
d Add correct state symbols after the formulae.

4 Write word equations, and then balanced equations, for:
a hydrochloric acid and ammonium hydroxide, reacting to form ammonium chloride and water
b hydrochloric acid and calcium hydroxide, reacting to form calcium chloride and water
c nitric acid reacting with sodium hydroxide
d a reaction between an acid and alkali which forms potassium nitrate and water

31 BASES

A base is a substance which neutralises acids. Alkalis are one sort of base. An alkali is a base that is soluble in water.

A base is a substance which can neutralise an acid

There are many different substances which can neutralise acids. They include:

metal hydroxides e.g. sodium hydroxide
metal oxides e.g. copper oxide
metal carbonates e.g. zinc carbonate
metal hydrogencarbonates e.g. sodium hydrogencarbonate
ammonia solution
metals e.g. magnesium

All of these substances are **bases**. A base is a substance which can neutralise an acid.

Some bases are also **alkalis**. An alkali is a base which is soluble in water.

> ### DID YOU KNOW?
> Magnesium oxide is a base which is used to neutralise acids in the stomach. It is not absorbed before it can do this. It can easily raise the pH of the stomach by about one unit.

Bases neutralise acids by removing H⁺ ions

You have seen how metal hydroxides, such as sodium hydroxide, neutralise acids. The H⁺ ions – which are what make a solution acidic – combine with the OH⁻ ions from the hydroxide. Water is formed.

$$H^+(aq) \quad + \quad OH^-(aq) \quad \rightarrow \quad H_2O(l)$$

| the acid part of the acid | the alkaline part of the alkali | a neutral compound |

Bases neutralise acids by removing the hydrogen ions from solution. The following examples show how each of the six kinds of bases listed above neutralise acids.

Metal hydroxide + acid → metal salt + water

You have met this one before. An example is the neutralisation of hydrochloric acid by sodium hydroxide. Sodium hydroxide is a soluble base, or alkali.

hydrochloric acid + sodium hydroxide →
sodium chloride + water

$$HCl(l) + NaOH(aq) \rightarrow NaCl(aq) + H_2O(l)$$

INVESTIGATION 31.1

Metal hydroxide + acid → metal salt + water

An example of this is the reaction between hydrochloric acid and sodium hydroxide, forming sodium chloride and water.

1 Fill a burette to zero with hydrochloric acid.
2 Use a pipette and filler to place 25 cm³ sodium hydroxide solution in a conical flask. Add three drops of the indicator phenolphthalein, which will turn it pink.
3 Slowly add the acid to the alkali, swirling all the time, until the indicator turns completely colourless. This is called the **end-point**. The procedure you have just carried out is called a **titration.**
4 Now that you know roughly where the end-point is, repeat steps 2 and 3. This time, take more care as you near the end-point, so that you get a more accurate value of the amount of acid needed to neutralise the alkali.
5 You have now found out exactly how much acid neutralises 25 cm³ of alkali. Your flask contains salt solution, contaminated with a little indicator. If you want to get pure salt crystals, you now need to repeat the titration without adding indicator. Use 25 cm³ of alkali, and the amount of acid you know is needed to neutralise it.
6 You now have a solution of sodium chloride. To speed up the formation of the salt crystals, boil away approximately 90 % of the water and leave to stand.

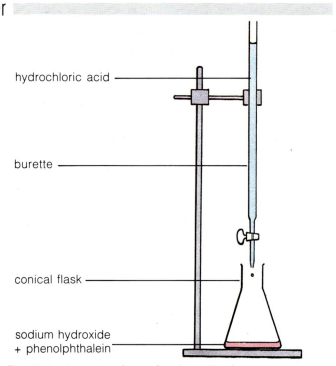

hydrochloric acid
burette
conical flask
sodium hydroxide + phenolphthalein

Fig. 31.1 Apparatus for performing a titration

Ammonia solution + acid → ammonium salt + water

When ammonia gas dissolves in water, some ammonium hydroxide forms.

$$NH_3(aq) + H_2O(l) \rightarrow NH_4OH(aq)$$

So the following reaction is really the same as the reaction between a metal hydroxide and an acid:

nitric acid + ammonium hydroxide → ammonium nitrate + water

$$HNO_3(aq) + NH_4OH(aq) \rightarrow NH_4NO_3(aq) + H_2O(l)$$

The salt formed in this reaction is ammonium nitrate. This is sold as fertiliser. As you can see from its formula, it contains a lot of nitrogen. Nitrogen is essential for plants to make proteins. Ammonium nitrate is the most widely used fertiliser in the world.

INVESTIGATION 31.2

You can carry out this experiment in just the same way as Investigation 31.1, but using ammonium hydroxide solution instead of sodium hydroxide solution. You will also need to use a different indicator. Use screened methyl orange instead of phenolphthalein. It will be green when you add it to the alkali, but turns grey when you reach the end point of the reaction.

Questions

1   Is ammonia an alkali?

2   Write word equations, and then balanced molecular equations, for the reactions which would occur between each of the following:
 a ammonia solution and hydrochloric acid
 b ammonia solution and sulphuric acid

EXTENSION

Fig. 31.2   Screened methyl orange in, from left, alkaline, neutral and acidic solutions.

Metal oxide + acid → metal salt + water

Metal oxides produce oxide ions, O^{2-}. The oxide ions react with the H^+ ions in acids, to form water.

$$2H^+ + O^{2-} \rightarrow H_2O$$

An example of such a reaction is the neutralisation of sulphuric acid by zinc oxide:

sulphuric acid + zinc oxide → zinc sulphate + water

$$H_2SO_4(aq) + ZnO(s) \rightarrow ZnSO_4(aq) + H_2O(l)$$

Notice that the zinc oxide has the symbol (s) after it. This means that it is in solid form, not dissolved in water. Zinc oxide is not soluble in water. It is a base, but not an alkali.

Questions

1   Is zinc sulphate soluble in water?

2   Write word equations, and then balanced molecular equations, for the reactions which would occur between each of the following:
 a hydrochloric acid and zinc oxide
 b sulphuric acid and copper (II) oxide
 c sulphuric acid and magnesium oxide

EXTENSION

INVESTIGATION 31.3

Metal oxide + acid → metal salt + water

You are going to add copper (II) oxide to dilute sulphuric acid, to form copper sulphate and water.

1   Put about $25\ cm^3$ of dilute sulphuric acid into a beaker. Warm the acid *gently*. This will speed up the reaction when you add copper oxide to it. When the acid is warm, put the beaker on the bench.

2   Add one spatula full of copper (II) oxide to the warm acid. Stir thoroughly. When all the copper (II) oxide has dissolved, add another spatula full.

3   Every now and then, test with universal indicator paper. When the copper (II) oxide seems *very* slow to dissolve, you are reaching the end of the reaction. Stop when no more copper (II) oxide will dissolve and the pH is nearly 7.

4   If you want a sample of the copper sulphate you have made, first filter the solution, to get rid of any copper (II) oxide that did not react. You now have a copper sulphate solution. Concentrate it by boiling off excess water and leave it to crystallise.

Metal carbonate or hydrogencarbonate + acid → metal salt + water + carbon dioxide

This reaction is often used to find out if an unknown substance is a carbonate or hydrogencarbonate. The carbon dioxide being given off in the reaction is usually very obvious. The solution fizzes (**effervesces**) as the gas is given off. The gas can be tested by passing it into lime water, which goes milky if it really is carbon dioxide.

An example of this reaction is the neutralisation of hydrochloric acid by calcium carbonate:

hydrochloric acid + calcium carbonate → calcium chloride + water + carbon dioxide

$$2HCl(aq) + CaCO_3(s) \rightarrow CaCl_2(aq) + H_2O(l) + CO_2(g)$$

Another example is the neutralisation of hydrochloric acid by sodium hydrogencarbonate. This reaction can happen in your stomach! If you have indigestion, it sometimes helps to swallow a spoonful of sodium hydrogencarbonate (baking soda). This neutralises the hydrochloric acid in your stomach. What do you think happens to the carbon dioxide produced?

hydrochloric acid + sodium hydrogencarbonate → sodium chloride + water + carbon dioxide

$$HCl(l) + NaHCO_3(aq) \rightarrow NaCl(aq) + H_2O(l) + CO_2(g)$$

Questions

1 Is calcium carbonate a base?
2 Is calcium carbonate an alkali?

3 Write word equations and then balanced molecular equations, for the reactions which would occur between each of the following:
a sulphuric acid and zinc carbonate
b nitric acid and potassium carbonate
c hydrochloric acid and potassium hydrogencarbonate
4 You are given a white powder, which you think might be a carbonate or hydrogencarbonate. How could you find out? What happens if you are right?
5 Many indigestion tablets contain sodium hydrogencarbonate.
a Why is this?
b Given three different types of indigestion tablets, what experiments could you do to see which acts fastest (without swallowing them!)?

EXTENSION

Fig. 31.3 Spreading lime on to agricultural land. The lime could be calcium hydroxide, or calcium carbonate. These are bases, and will reduce acidity in soil. Lime also improves the texture of heavy soils.

INVESTIGATION 31.4

Metal carbonate + acid → metal salt + water + carbon dioxide

In this reaction, you will use magnesium carbonate and dilute sulphuric acid. The reaction is:

magnesium carbonate + sulphuric acid → magnesium sulphate + water + carbon dioxide

$$MgCO_3(s) + H_2SO_4(aq) \rightarrow MgSO_4(aq) + H_2O(l) + CO_2(g)$$

1 Put about 25 cm³ of dilute sulphuric acid into a beaker.
2 Add one spatula measure of magnesium carbonate. It will seem to disappear as it reacts. The solution will fizz, as carbon dioxide gas is produced. (What do you think happens to the water that is produced? What happens to the magnesium sulphate which is produced?)

3 Continue adding magnesium carbonate until there is no more fizzing, and the magnesium carbonate does not disappear. The acid is now neutralised.
4 If you want a sample of the magnesium sulphate that you have made, do the following. First, take your beaker of neutralised acid – which now contains your magnesium sulphate solution. Filter it, to remove any magnesium carbonate which did not react. Then concentrate the solution, and leave to crystallise.

Metal + acid → metal salt + hydrogen

Many metals react with acids, neutralising them. In this case, the hydrogen ions do not combine with oxygen to form water. The hydrogen ions from the acid gain electrons from the metal atoms. Then the hydrogen atoms join in pairs to form hydrogen gas:

$$2H^+(aq) + 2e^- \rightarrow H_2(g)$$

The acid fizzes as the hydrogen bubbles out. You can show that the gas is hydrogen by putting a lighted splint into the test tube. If the gas is hydrogen, it gives a squeaky pop.

Some metals, like silver, do not react with acids. Others, such as sodium, react dangerously fast. Zinc, iron and magnesium all react at a steady speed.

zinc + sulphuric acid → zinc sulphate + hydrogen

$$Zn(s) + H_2SO_4(aq) \rightarrow ZnSO_4(aq) + H_2(g)$$

Questions

1   Write word equations and then balanced molecular equations, for the reactions which occur between:
    a magnesium and hydrochloric acid
    b iron and sulphuric acid (the iron compound formed is pale blue-green)
2   When metals react with nitric acid, different reactions may take place. Find out about these.

— *EXTENSION* —

INVESTIGATION 31.5

Metal + acid → metal salt + hydrogen

In this experiment you will use magnesium ribbon and dilute hydrochloric acid. The reaction is:

magnesium + hydrochloric → magnesium chloride
             acid   + hydrogen

1   Put about 25 cm³ of dilute hydrochloric acid into a beaker.
2   Add a strip of magnesium ribbon. The solution will fizz, as hydrogen gas is given off. You can test the gas with a burning splint if you like.
3   Continue adding magnesium ribbon, bit by bit, until the fizzing stops. The acid has now been neutralised.

Questions

1   What can you now do if you want to obtain a sample of the magnesium chloride you have made?
2   Do you think that it would make any difference to the speed of this reaction if you used magnesium powder instead of ribbon? Design an experiment to find out the answer to this question.

Summary – Learn these!

acid + metal → metal salt + hydrogen

acid + metal oxide → metal salt + water

acid + metal hydroxide → metal salt + water

acid + metal carbonate → metal salt + water + carbon dioxide

acid + metal hydrogencarbonate → metal salt + water + carbon dioxide

acid + ammonia solution → ammonium salt + water

More summary – Remember these!

When **hydrochloric** acid is neutralised, the salt formed is a metal **chloride**.

When **nitric** acid is neutralised, the salt formed is a metal **nitrate**.

When **sulphuric** acid is neutralised, the salt formed is a metal **sulphate**.

32 ACID RAIN

Acid rain is caused when sulphur dioxide and nitrogen oxide gases are released into the air. They dissolve in water droplets, and fall to earth as rain or snow.

Sulphur dioxide is emitted when fossil fuels are burnt

Fossil fuels, such as coal, natural gas and oil, contain sulphur. When they are burnt, the sulphur combines with oxygen. It forms sulphur dioxide. The sulphur dioxide is given off as a gas.

Sulphur dioxide is an unpleasant gas. It damages living things. Humans who breathe in a lot of sulphur dioxide over a long period of time have an increased risk of suffering from colds, bronchitis and asthma. Sulphur dioxide can kill plant leaves. It may completely kill the plant.

Fig. 32.1   The upper branches of this tree may have been killed by acid rain.

INVESTIGATION 32.1

The effect of sulphur dioxide on plants

Sodium metabisulphite solution gives off sulphur dioxide. You are going to test its effect on two kinds of important crop plants.

1   Fill four containers with damp compost. Press the compost down firmly, but not too hard. Make the tops level.

2   Put barley seeds on to the compost in two containers. Spread them out evenly. Put maize seeds on to the compost in the other two containers.

3   Cover all the seeds with enough compost to hide them completely. Label all four containers.

4   Leave the containers in a warm place until most of the seeds have germinated. Keep them well watered.

5   When the shoots of barley and maize are about 3–5 cm tall,

make labelled diagrams of each of the four sets of seedlings.

6   Now take four small containers, such as watch glasses. Into two of them place a piece of cotton wool soaked in sodium metabisulphite solution. (TAKE CARE. Do not get it on your hands.) In the other two, place a piece of cotton wool soaked in water.

7   Place each container of seedlings in a large plastic bag, with one of the watch glasses. You should have one of your groups of barley seedlings with a container of sodium metabisulphite solution and one with a container of water. Do the same with the maize seedlings. Tie the plastic bags tightly, so that no air can get in or out.

8   Make labelled diagrams of each of your four sets of seedlings about

30 min after putting them into their bags, and again after one or two days.

Questions

1   Why were some seedlings enclosed with sodium metabisulphite solution, and some with water?

2   Were both types of plant affected in the same way by sulphur dioxide? Explain what these effects were, and any differences between the maize and barley seedlings.

Sulphur dioxide and nitrogen oxides form acid rain

Sulphur dioxide dissolves in water to form an acid solution. Nitrogen oxides also form an acid solution when they dissolve. Nitrogen oxides form when the nitrogen and oxygen in the air react together under the high temperatures and pressures inside a petrol engine. Sulphur dioxide and nitrogen oxides are carried high into the air. Here, they dissolve in water droplets in clouds. They make the water droplets acid. The clouds may be carried many miles before they drop their water. It falls as acid rain or snow.

Acid rain damages trees and aquatic animals

Acid rain can damage trees. Acid rain washes important minerals, such as calcium and magnesium, out of the soil. Acid rain falling on thin soils, such as those in the mountains of Scandinavia, can kill huge areas of forest.

As the acid water runs through the soil, it washes out aluminium ions. The aluminium runs into rivers and lakes. Aluminium is very toxic to fish. Some acid rivers and lakes now contain hardly any fish.

Acid rain damages buildings

Acid rain can also damage buildings. The acid reacts with carbonates in limestone. The limestone dissolves, so that the stone gradually crumbles away.

Fig. 32.2   Stonework in Lincoln Cathedral

What is the major cause of acid rain?

Normal rain is slightly acid anyway. This is because carbon dioxide in the air dissolves in rain drops to form carbonic acid. Carbonic acid is a very weak acid. So even ordinary rain has a pH a little below 7.

But the large amount of sulphur dioxide and nitrogen oxides which we are now releasing into the air are making rain much more acid than this. We are damaging plants, animals and buildings. Something must be done to stop this getting any worse.

Scientists are still not sure about the most important cause of acid rain. Is it power stations burning coal? Certainly these are very important. Large amounts of sulphur dioxide are released from these power stations. Many of them are now beginning to remove the sulphur dioxide from the smoke they produce, so that it does not go into the air. But this is expensive and means that we will have to pay more for the electricity they make. Or are cars the most important producers of acid rain? Car exhaust fumes contain nitrogen oxides. Since 1956 the amounts of sulphur dioxide emitted by coal burning have gone down. But the acid rain problem has got worse! Over this time the numbers of cars on the roads has increased enormously. So it looks as though cars are as much to blame as coal-burning factories and power stations.

Fig. 32.3   Acid rain

Combustion of fossil fuels, for example oil or coal in power stations and petrol in cars, releases oxides of sulphur and nitrogen into the air.

The sulphur and nitrogen oxides dissolve in water droplets in clouds, making them acidic. They may be carried for hundreds of kilometres.

Plants may be damaged when acid rain falls on them. Acidification of the soil allows toxic elements to dissolve, and be washed into streams, rivers and lakes. Fish and other aquatic organisms may be killed.

Questions

1 a How is sulphur dioxide produced?
  b What damage can be done by sulphur dioxide gas?
  c What is produced when sulphur dioxide dissolves in water?
2 Both sulphur dioxide and nitrogen oxides contribute to the formation of acid rain.
  a What are the major sources of these two gases?
  b Describe the kind of damage which can be done by acid rain.
  c Is normal rain neutral, acidic or alkaline? Explain your answer.
  d What do you think could be done to reduce the damage caused by acid rain?

Questions

1 Copy and complete this table using the suggestions below.

Type of substance		strong acid		neutral	
pH	5				14
Colour with universal indicator			green/blue		

Choose from: 2, yellow, weak acid, weak alkali, purple, green, 7, 8, strong alkali, red.

2 Copy and complete this passage:

Acids are solutions containing ..... ions. Alkalis are solutions containing ..... ions. If an acid and an alkali are mixed, the two types of ion combine to form ..... This process is called ..... .

3 Explain the difference between:

a a strong acid and a concentrated acid.

b a weak acid and a dilute acid.

4 Nitric acid is neutralised by calcium carbonate.

a What three products are formed in this reaction?

b Write a word equation for this reaction.

c Write a molecular equation for this reaction.

d Balance your equation and add state symbols.

e Describe how you could use this reaction to obtain a sample of a salt.

---

5 Repeat question 4 for the following reactions:
a nitric acid and lead carbonate
b hydrochloric acid and copper oxide
c sulphuric acid and zinc
6 The salt sodium nitrate can be made by titration.
a Which reactants would you use?
b Write a word equation for the reaction.
c Write a balanced molecular equation for the reaction.
d Describe fully how you would make the salt, explaining why you would titrate three times.

— E X T E N S I O N —

DID YOU KNOW?

Amounts of sulphur dioxide being emitted in the United Kingdom are decreasing. In 1980, 4.67 million tonnes were emitted, but this had dropped to 3.74 million by 1986.

7 The table shows the pH range at which certain plants will grow successfully.

Plant	pH range
apples	5.5 - 7.0
blackcurrants	6.0 - 7.5
broad beans	5.5 - 7.0
carrots	6.0 - 7.5
lettuces	6.5 - 7.5
onions	6.5 - 7.5
potatoes	5.5 - 6.5
tomatoes	5.5 - 7.0

a Design a chart or graph to show this information as clearly as possible.

b A market gardener wishes to grow as wide a range of crops as possible. He decides to test the pH of his soil. Describe how he could do this.

c He finds that his soil's pH is 6.0. Is the soil acid or alkaline?

d Which of the above crops could he grow successfully?

e To grow a fuller range of crops, he would like the pH of his soil to be around 6.5. Which crops could he grow then?

f He decides to add lime to the soil. Lime can be calcium carbonate. What effect will this have?

8 Describe how you would prove the following:

a Magnesium and hydrochloric acid react to produce hydrogen.

b Magnesium carbonate and hydrochloric acid react to produce carbon dioxide.

c Sodium hydroxide solution will neutralise vinegar.

d Cola drinks contain stronger acid than orange squash.

e Rhododendrons grow best in acid soils.

9 Explain the following:

a Rubbing sodium bicarbonate on a bee sting helps to stop the pain.

b Rubbing vinegar on a wasp sting helps to stop the pain.

c Lakes in limestone areas are much less affected by acid rain.

d Acid snow affects the young of water animals more than acid rain affects them.

REVERSING THE EFFECTS OF ACID RAIN

Much of the sulphur dioxide and nitrogen oxides which cause acid rain are produced by coal-burning power stations. Efforts are now being made to reduce the amount of acid gases which they produce.

In 1988 the Central Electricity Generating Board in Great Britain was given the go-ahead to build the world's biggest sulphur removal plant at the Drax power station in North Yorkshire. The plant cost £400 million and created 600 construction jobs. It cut Drax's sulphur dioxide emissions by 90%.

The sulphur dioxide is washed out of the gases emitted from the power station by spraying a watery slurry of limestone over the gas. The sulphur dioxide reacts with the limestone to produce gypsum (calcium sulphate). The gypsum can be used for making plasterboard.

The cost of this plant, and others to follow, has to be passed on to consumers in higher electricity prices.

Scientists are very pleased that these sulphur removal plants are being introduced, but are concerned that damage already done by acid rain may be irreversible. A group of environmental scientists in Norway is conducting an experiment to find out if badly damaged soils can recover from acid rainfall. They are investigating the effects of stopping acid rain falling on an area in southern Norway. The area receives rain and snow with a pH of 4.2. It has a thin soil with a sparse covering of pine and birch trees. They have divided their experimental area into three parts. One part, area A, has been left alone. The other two areas, B and C, have been covered with plastic roofs. All the rain falling on the roofs is collected. In area B, the rain which falls on to the roof has the acidity removed from it. It is then sprinkled on to the trees growing under the roof. In area C, the rain which falls on to the roof is sprinkled on to the trees without having its acidity removed.

The scientists have found that in areas A and C the soil has remained very acidic. The water running off from these areas contains a lot of nitrate and sulphate ions. But in area B the amount of nitrate in the runoff water went down by 60% within two weeks of the beginning of the experiment. After three years the sulphate levels in the runoff water had gone down by 50%.

The researchers think that these results are encouraging. They show that the effects of acid rain can be reversed if we stop releasing sulphur dioxide and nitrogen oxide into the air. But it will take a long time for new trees to grow in areas where they have been killed, and for new populations of fish to build up in lakes damaged by acid runoff.

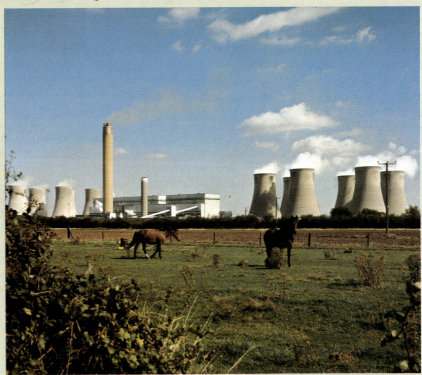

Drax power station before the sulphur removal plant was built. The white clouds from the cooling towers are just water vapour. More harmful gases, such as sulphur dioxide, are coming from the tall, narrow tower.

a Why do you think that the CEGB took so long to introduce sulphur removal plants at its coal-burning power stations?

b Explain how sulphur dioxide can be removed from the waste gases emitted by coal-burning power stations.

c Name one useful by-product from this process.

d One scientist said, of the decision to build the sulphur removal plant at Drax, 'It's welcome, but it's a bit like closing the stable door after the horse has bolted.' What do you think he meant?

e In the experiment in Norway, why did the scientists cover part of their experimental area with plastic roofs?

f The two areas covered with plastic roofs were treated differently. Explain why you think they used these two methods of treatment.

g Briefly summarise the results of the Norwegian experiment, and explain why they are encouraging.

33 EXOTHERMIC AND ENDOTHERMIC REACTIONS

Exothermic reactions give out heat.
Endothermic reactions take in heat.

The reaction between magnesium and sulphuric acid

You will probably have seen how magnesium reacts with sulphuric acid.

The mixture fizzes as hydrogen is given off. The magnesium disappears. The equation for this reaction is:

magnesium + sulphuric acid → magnesium sulphate + hydrogen

$$Mg(s) + H_2SO_4(aq) \rightarrow MgSO_4(aq) + H_2(g)$$

This reaction shows an important property of acids. Acids corrode metals. But there are also two other things happening which have great importance to chemists.

1 *Heat is produced* If you measure the temperature of the acid before and after adding the magnesium, you will find that the temperature goes up.

2 *Electrons move from particle to particle* Before it reacts, the magnesium is in the form of uncharged magnesium atoms. The sulphuric acid is made up of hydrogen ions with a single plus charge – H^+ – and sulphate ions with a double minus charge – SO_4^{2-}.

During the reaction, the magnesium atoms each lose two electrons. They become ions with two plus charges – Mg^{2+}. The electrons are passed to the H^+ ions. They become uncharged H atoms, which join together in pairs to form H_2 molecules.

Chemists are particularly interested in changes in temperature and in electron movement during chemical reactions.

Fig. 33.1 Magnesium reacts vigorously with sulphuric acid, releasing bubbles of hydrogen.

Many reactions give out heat

Most chemical reactions are like the one between magnesium and sulphuric acid. They give out heat to the surroundings. Reactions which give out heat are called **exothermic** reactions. All the reactions we call 'burning' are exothermic reactions. Some examples include:

natural gas burning: $CH_4(g) + 2O_2(g) \rightarrow CO_2(g) + 2H_2O(l)$

coal burning: $C(s) + O_2(g) \rightarrow CO_2(g)$

magnesium burning: $2Mg(s) + O_2(g) \rightarrow 2MgO(s)$

Another very similar reaction occurs in your cells. It is called respiration. It, too, is an exothermic reaction:

respiration: $C_6H_{12}O_6(aq) + 6O_2(aq) \rightarrow 6CO_2(aq) + 6H_2O(l)$

There are many other exothermic reactions which you will be familiar with. For example, as vegetable matter rots on a compost heap, heat is given out. The reactions involved are exothermic. Another exothermic reaction which you have probably performed in the laboratory is the neutralisation of an acid by an alkali.

Fig. 33.2 A flare-out at an oil refinery. This oxidation reaction is exothermic.

Some chemical reactions take in heat

In most chemical reactions, heat is given out. But there are some reactions which take in heat from their surroundings. Reactions which take in heat are called **endothermic reactions**.

One endothermic chemical reaction which is important in industry is the decomposition of calcium carbonate. Calcium carbonate is limestone or chalk. The reaction is:

calcium carbonate → calcium oxide + carbon dioxide

$$CaCO_3(s) \rightarrow CaO(s) + CO_2(g)$$

'Decomposition' means 'breaking down'. A decomposition reaction is one in which a substance breaks down into two or more new substances. In this reaction the calcium carbonate breaks down, or decomposes, to calcium oxide and carbon dioxide. Calcium oxide is called lime, or **quicklime**. It has many uses. Perhaps the most important is in the manufacture of cement. It is also spread on acidic soils to increase their pH, and on clay soils to improve their structure. This decomposition reaction is endothermic. When 1 kg of calcium carbonate decomposes, nearly 3000 kJ of heat are taken in! So the reaction will not happen unless a lot of heat is supplied to it. This is done by putting the calcium carbonate into large ovens called lime kilns. They are kept at around 800 °C.

Fig. 33.4   Limestone is decomposed to form calcium oxide, an important component of cement. The bricklayer is using mortar, a mixture of cement, sand and water.

Another important industrial reaction which is endothermic is the production of sodium and chlorine from salt. This is also a decomposition reaction.

sodium chloride → sodium + chlorine

$$2NaCl(l) \rightarrow 2Na(l) + Cl_2(g)$$

The sodium produced in this reaction has many uses. These include making soap and cooling nuclear power stations. The chlorine can be used to make PVC (polyvinyl chloride) and bleach.

Fig. 33.3   The combustion of magnesium is an exothermic reaction.

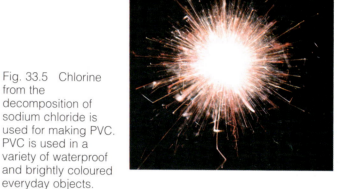

Fig. 33.5   Chlorine from the decomposition of sodium chloride is used for making PVC. PVC is used in a variety of waterproof and brightly coloured everyday objects.

Questions

1 a What is meant by an exothermic reaction?
  b Give one example of an exothermic reaction.
  c What is meant by an endothermic reaction?
  d Give one example of an endothermic reaction.
  e Which type of reaction is most common – exothermic or endothermic?
2 a What is lime?
  b What is lime used for?
  c How is lime produced?

  d Is the reaction which produces lime an exothermic or endothermic reaction?
  e What will be the chief costs of making lime?
  f Lime-making involves a decomposition reaction. Give another example of a decomposition reaction, and explain what decomposition means.

34 MORE ABOUT EXOTHERMIC AND ENDOTHERMIC REACTIONS

INVESTIGATION 34.1

Exothermic and endothermic changes

Carry out the following five experiments. For each one, measure and record the temperature of the liquid. Then add the solid and stir. Finally, measure the temperature of the liquid again.

1 Add two spatula measures of fine iron filings to 25 cm³ of copper sulphate solution.
2 Add two spatula measures of potassium sulphate to 25 cm³ of water. *Mg SO₄*
3 Add two spatula measures of ~~ammonium nitrate~~ to 25 cm³ of water.
4 Add a strip of magnesium ribbon to 25 cm³ of hydrochloric acid.
5 Add two spatula measures of anhydrous copper sulphate to 25 cm³ of water.

6. *Conc H₂SO₄ → sugar - amo.*

Questions

1 Which reactions were exothermic?
2 Which reactions were endothermic?

Bond making is exothermic

When you add water to calcium oxide, the particles in the water and the particles in the calcium oxide form bonds. The bonds hold the five atoms together as a molecule of calcium hydroxide.

$$CaO(s) + H_2O(l) \rightarrow Ca(OH)_2(s)$$

As the bonds are made, heat is released into the surroundings. This reaction is exothermic. Bond making is exothermic.

Bond breaking is endothermic

Calcium carbonate molecules are made up of five atoms. (What are they?) When calcium carbonate is heated, these molecules break up to form a two atom molecule, CaO, and a three atom molecule, CO_2.

$$CaCO_3(s) \rightarrow CaO(s) + CO_2(g)$$

So bonds are broken in this reaction. Bond breaking takes in heat from the surroundings. So this reaction is endothermic. Bond breaking is endothermic.

Decomposition involves breaking bonds. So decomposition reactions are usually endothermic. Some examples include:

$$2NaNO_3(s) \rightarrow 2NaNO_2(s) + O_2(g)$$

$$NH_4Cl(s) \rightarrow NH_3(g) + HCl(g)$$

$$2HgO(s) \rightarrow 2Hg(l) + O_2(g)$$

EXTENSION

Some reactions are exothermic, while others are endothermic

Chemical reactions can be thought of as happening in two stages. (i) Firstly the bonds in the molecules of the reactants have to be broken, forming free atoms. (ii) Secondly these free atoms bond together to form the product molecules. The first stage is endothermic, and the second stage is exothermic. Whether or not the overall energy change is endothermic or exothermic depends on which of the two stages has the largest energy value.

For example, when natural gas (methane) burns, the two stages can be represented by the **energy level diagram** shown in Figure 34.1a. Stage (i) involves the breaking of the bonds in the methane molecule and the two oxygen molecules while stage (ii) involves the bonding together of the nine atoms to form a carbon dioxide molecule and two water molecules. The value for stage (ii) is bigger than the value for stage (i), so the overall energy change is **exothermic**.

When ammonia breaks down to give nitrogen and hydrogen the two stages can be represented by the energy level diagram shown in figure 34.1b. Stage (i) involves the breaking of bonds in two ammonia molecules while stage (ii) involves the bonding together of the eight atoms to form one nitrogen molecule and three hydrogen molecules. The value for stage (i) is bigger than the value for stage (ii), so the overall energy change is **endothermic**.

As these two energy level diagrams suggest, in many chemical reactions the first stage of the reaction is endothermic, so energy is needed in order to get the reaction started. When molecules of the reacting substances meet each other they must have a certain amount of energy or they won't react. This energy amount is called the **activation energy** of the reaction. Reactions with a small activation energy tend to go quickly (because lots of the molecules in the mixture can react), but reactions with a large activation energy tend to go slowly (because few of the molecules in the mixture can react). Scientists have discovered certain substances that, when added to a reaction mixture, decrease the energy values of stages (i) and (ii) by equal amounts. The **overall** energy change for the reaction is therefore not changed, but because the activation energy is decreased the reaction happens more quickly. Substances that do this are called **catalysts**.

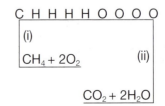

Fig. 34.1a

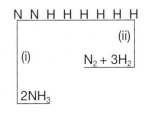

Fig. 34.1b

88

Questions

1 The energy values of some bonds are as follows:
C–H 412, O=O 496, O–H 463, and C=O 743.

a Work out the energy value of stage (i) for the reaction between methane and oxygen. To do this, add together the energy values of all the bonds in one methane molecule and two oxygen molecules.

b Work out the energy value of stage (ii) for the reaction between methane and oxygen. To do this, add together the energy values of all the bonds in one carbon dioxide molecule and two water molecules.

c Work out the overall energy value for the burning of methane by finding the difference between your value for stage (i) and your value for stage (ii). Will this reaction be exothermic or endothermic?

2 The energy values of some bonds are as follows:
N≡N 944, H–H 436, and N–H 388.

a Work out the energy value of stage (i) for the decomposition of ammonia.

b Work out the energy value of stage (ii) for the decomposition of ammonia.

c Work out the overall energy value for the decomposition of ammonia. Will this reaction be exothermic or endothermic?

A substance which can undergo an exothermic reaction has chemical energy

Stored energy, or potential energy, can be in many different forms. One form of potential energy is chemical energy. A substance which can undergo an exothermic reaction has chemical energy. Petrol has chemical energy. When it burns in an engine, this stored energy is transferred to heat energy and kinetic (movement) energy. The burning of petrol is an exothermic reaction.

In an exothermic reaction, the energy released is usually released as heat. But humans have built machines which cause the energy to be released in other useful forms. A car engine is a good example. If petrol is poured on the ground and set alight, the exothermic reaction produces heat, which disperses into the ground and air. But in a car engine the petrol burns in the cylinder. The hot, fast moving gas molecules which are produced push the piston, which turns the crankshaft. So some of the stored chemical energy in the petrol is transferred to kinetic energy of the car.

chemical energy → heat energy → kinetic energy
  (in the fuel)   (as the   (as the car
               fuel burns)   moves)

In a coal-fired power station, coal burns in an exothermic reaction. The heat produced is used to boil water. The steam produced turns turbines. The turbines turn generators, which generate (produce) electricity.

chemical → internal → kinetic → electrical
energy   energy   energy   energy

(in coal)   (in water   (in turbines
          and steam)   and generator)

In your cells, glucose from food combines with oxygen. This is an exothermic reaction. Some of the energy is released as heat. This keeps your body temperature at 37 °C, even when the temperature of your environment is much less. But some of the energy is used for other activities, such as movement.

chemical energy ⟶ movement   and   heat
  (from food)   (in muscles)   (from muscles)

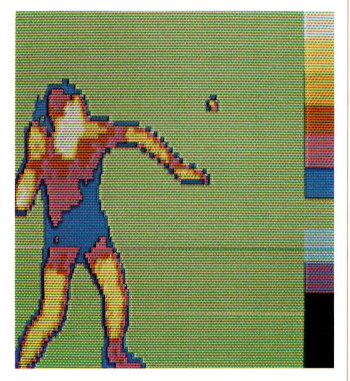

Fig. 34.2 Respiration is an exothermic reaction, which releases heat energy. The heat image, or thermogram, of a man playing squash shows the relative temperatures of different parts of his body. The scale at the side runs from hot (white) at the top to cold (black) at the bottom. Notice that the squash player is hotter than his surroundings. (Why do you think the squash ball is hot?)

EXTENSION

35 OXIDATION AND REDUCTION

Many chemical reactions involve movement of electrons.

Chemical reactions often involve movement of electrons

Think again about the reaction between magnesium and sulphuric acid:

magnesium + sulphuric acid → magnesium sulphate + hydrogen

$$Mg(s) + H_2SO_4(aq) \rightarrow MgSO_4(aq) + H_2(g)$$

This equation gives an overall picture of the reaction. But if we include *charges* in the equation, we can see even more about what is happening.

$$Mg(s) + H^+_2SO_4{}^{2-}(aq) \rightarrow Mg^{2+}SO_4{}^{2-}(aq) + H_2(g)$$

Look first at what is happening to the magnesium

particles. At the beginning of the reaction, they are uncharged magnesium atoms, Mg(s). But at the end of the reaction, they have become positively charged magnesium ions, Mg^{2+}. The magnesium particles have lost electrons.

Where have these electrons gone? Look at the hydrogen particles. At the beginning of the reaction, they are positively charged hydrogen ions, H^+. At the end of the reaction, they are uncharged hydrogen atoms bonded together in pairs to form hydrogen molecules, H_2. Each hydrogen ion has gained an electron.

So during this reaction, electrons move from magnesium particles to hydrogen particles. Many chemical reactions involve similar electron movements.

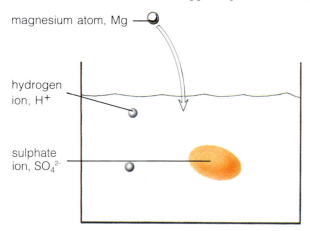

magnesium atom, Mg

hydrogen ion, H^+

sulphate ion, $SO_4{}^{2-}$

Fig. 35.1 The reaction between magnesium and sulphuric acid $H_2SO_4(aq)$. The water molecules in the sulphuric acid solution are not shown.
Hydrogen ions have a single positive charge; they have one more proton than electrons. Magnesium atoms have no overall charge; they have equal numbers of electrons and protons.

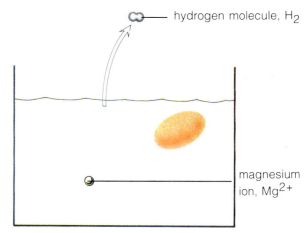

hydrogen molecule, H_2

magnesium ion, Mg^{2+}

When a magnesium atom meets two hydrogen ions in solution, each hydrogen ion takes one electron from the magnesium atom. This leaves the magnesium with a double positive charge. The two hydrogen atoms, now with no net charge, join together to form a molecule of hydrogen.

Loss of electrons is called oxidation

Another example of a chemical reaction which involves movement of electrons is the burning of magnesium.

magnesium + oxygen → magnesium oxide

$$2Mg(s) + O_2(g) \rightarrow 2MgO(s)$$

Putting in the charges, we get:

$$2Mg(s) + O_2(g) \rightarrow 2Mg^{2+}O^{2-}(s)$$

Again, the magnesium particles have lost electrons.
This happens with many metals when they react with oxygen. The metals lose electrons, which are transferred to the oxygen. We say that the metal has been **oxidised**.

Chemists do not only use the word 'oxidised' when oxygen is involved in the reaction. It is used whenever electrons are lost, even if they are transferred to something other than oxygen. **Oxidation is the loss of electrons**. Sometimes the electrons might be taken up by oxygen; sometimes it might be something else. Look back at the magnesium and sulphuric acid reaction. The magnesium particles lost electrons. So we say that the magnesium was oxidised.

Gain of electrons is called reduction

If something loses electrons in a chemical reaction, something else must gain them. When magnesium burns, the electrons from the magnesium are transferred to oxygen. We say that the oxygen is reduced. **Reduction is the gain of electrons**.

Reactions involving electron transfer are called redox reactions

If one substance loses electrons in a reaction, then something else must gain them. So if one substance is oxidised, something else is reduced. Reactions like this are called 'reduction-oxidation' reactions or **redox reactions** for short.

How can you tell if a reaction is a redox reaction? Is the reaction between zinc and hydrochloric acid a redox reaction? You need to follow this method to find out:

1 Write an equation for the reaction.

$$Zn(s) + 2HCl(aq) \rightarrow ZnCl_2(aq) + H_2(g)$$

2 Put charges into the equation.

$$Zn(s) + 2H^+Cl^-(aq) \rightarrow Zn^{2+}Cl^-_2(aq) + H_2(g)$$

3 Look at each type of particle individually.

$$Zn \rightarrow Zn^{2+}$$

$$2H^+ \rightarrow H_2$$

$$2Cl^- \rightarrow 2Cl^-$$

4 Decide whether anything loses or gains electrons.
Zn *loses* electrons. So zinc is *oxidised*.
H^+ *gains* electrons. So hydrogen is *reduced*.
So this reaction is a redox reaction.

Fig. 35.2 The reaction between zinc and hydrochloric acid. Which particles are losing electrons? Which are gaining them?

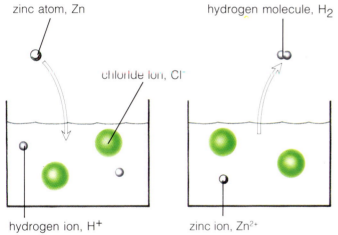

Zinc and copper sulphate solution (on the left) react together to form copper and zinc sulphate. You can see the copper forming in the bottom of the right-hand beaker.

Here is another example. Is the reaction between silver nitrate and hydrochloric acid a redox reaction?

1 Write the equation.

$$AgNO_3(aq) + HCl(aq) \rightarrow AgCl(s) + HNO_3(aq)$$

2 Put in charges.

$$Ag^+NO_3^-(aq) + H^+Cl^-(aq) \rightarrow Ag^+Cl^-(s) + H^+NO_3^-(aq)$$

3 Look at each type of particle individually.

$$Ag^+ \rightarrow Ag^+ \qquad\qquad H^+ \rightarrow H^+$$

$$NO_3^- \rightarrow NO_3^- \qquad\qquad Cl^- \rightarrow Cl^-$$

4 Decide whether anything loses or gains electrons.
Nothing does. So this is not a redox reaction.
Notice that a compound ion like the nitrate ion, NO_3^-, is considered as a single particle. You do not have to think about the nitrogen and oxygen atoms in it separately.

Fig. 35.3 As silver nitrate solution is added to hydrochloric acid, a white cloud of silver chloride immediately forms.

Questions

1 Which of the following are redox reactions?
a $Fe(s) + H_2SO_4(aq) \rightarrow FeSO_4(aq) + H_2(g)$
b $BaCl_2(aq) + H_2SO_4(aq) \rightarrow BaSO_4(s) + 2HCl(aq)$
c $Mg(s) + Cl_2(g) \rightarrow MgCl_2(s)$
d $Zn(s) + CuSO_4(aq) \rightarrow ZnSO_4(aq) + Cu(s)$

2 a Why is it that when a metal element reacts with a non-metal element, the reaction is always a redox reaction?

b Which element is oxidised – the metal or the non-metal?
c Which element is reduced – the metal or the non-metal?

3 Each of the following reactions is a redox reaction. For each one, say which substance is oxidised, and which is reduced.

a $4Na(s) + O_2(g) \rightarrow 2Na_2O(s)$
b $Mg(s) + ZnCl_2(aq) \rightarrow MgCl_2(aq) + Zn(s)$
c $2FeCl_2(s) + Cl_2(g) \rightarrow 2FeCl_3(s)$
d $2NaCl(s) \rightarrow 2Na(s) + Cl_2(g)$
e $Cl_2(g) + 2KI(aq) \rightarrow I_2(aq) + 2KCl(aq)$
f $TiCl_4(l) + 4Na(l) \rightarrow Ti(s) + 4NaCl(s)$

EXTENSION

36 SOME IMPORTANT REDOX REACTIONS

Redox reactions cannot always be identified by looking at electron movement.

The best definition of oxidation is 'the loss of electrons'. The best definition of reduction is 'the gain of electrons'. But in some reactions, these definitions can be hard to apply.

For example, when steam reacts with coal, the following reaction takes place:

$$H_2O(g) + C(s) \rightarrow CO(g) + H_2(g)$$

None of the particles involved in this reaction are ions. So there are no charges to put into the equation. Yet it *is* a redox reaction!

There is another definition of oxidation which can be used. **Oxidation is the gain of oxygen**. Similarly, **reduction is the loss of oxygen**. In this reaction, the steam loses oxygen. The steam is reduced. The coal (carbon) gains oxygen. The coal is oxidised.

There are other reactions where this definition is useful. For example, magnesium reacts with copper (II) oxide to give magnesium oxide and copper:

$$Mg(s) + CuO(s) \rightarrow MgO(s) + Cu(s)$$

Magnesium gains oxygen, so it is oxidised.
Copper loses oxygen, so it is reduced.

In this reaction, we could also use the first definition of oxidation and reduction. Putting charges into the equation, we get:

$$Mg(s) + Cu^{2+}O^{2-}(s) \rightarrow Mg^{2+}O^{2-}(s) + Cu(s)$$

Magnesium loses electrons, so it is oxidised.
Copper gains electrons, so it is reduced.

When you are deciding whether or not a reaction is a redox reaction, use both definitions. Firstly, ask if electrons are being gained or lost. Then ask if oxygen is being gained or lost. If the answer to *either* question is 'yes', then the reaction is a redox reaction. If the answer to *both* questions is 'no', then it is not a redox reaction.

Combustion of fuels

When a fuel burns, it gains oxygen. It is oxidised. Natural gas is a good example. Natural gas is methane, CH_4.

$$CH_4(g) + 2O_2(g) \rightarrow CO_2(g) + 2H_2O(g)$$

Fig. 36.1   Natural gas burning

Respiration

Respiration is the release of energy from food in living cells. Glucose from food reacts with oxygen. The carbon in glucose is oxidised.

glucose + oxygen → carbon dioxide + water

$$C_6H_{12}O_6(aq) + 6O_2(g) \rightarrow 6CO_2(g) + 6H_2O(l)$$

So we get energy from our food by a redox reaction.

Fig. 36.2   A redox reaction in the skier's cells releases energy for muscular movement.

Corrosion of metals

Many metals corrode or rust in air. Iron is a good example. The iron combines with oxygen. It is oxidised.

Iron + oxygen → iron oxide (rust)

$$4Fe(s) + 3O_2(g) \rightarrow 2Fe^{3+}_2O^{2-}_3(s)$$

The iron loses electrons, so it is oxidised. The oxygen gains electrons, so it is reduced. Rusting is a redox reaction.

Fig. 36.3   Rusting is a redox reaction.

Decomposing sodium chloride

Sodium chloride can be decomposed to sodium and chlorine.

$$2Na^+Cl^-(l) \rightarrow 2Na(l) + Cl_2(g)$$

The sodium ions gain electrons, so they are reduced. The chloride ions lose electrons, so they are oxidised. This decomposition is a redox reaction.

Making sodium hydroxide

A lot of the sodium which is produced from the decomposition of sodium chloride is used to make sodium hydroxide.

$$2Na(s) + 2H_2O(l) \rightarrow 2Na^+OH^-(aq) + H_2(g)$$

The sodium atoms lose electrons, so they are oxidised. The water molecules gain electrons, becoming hydroxide ions and hydrogen molecules, so the water molecules are reduced.

Redox reactions

Because scientists can understand redox reactions they have been able to use them in a variety of applications. They can create tiny power cells for quartz watches, control the rusting of iron and explain the role of oxygen in keeping us alive.

Fig. 36.5   Sodium reacting with water. Universal indicator solution has been added to the water. It was green to begin with, but is turning purple as the sodium and water react. Why?

Extraction of metals from ores

Iron is extracted from iron ore in a blast furnace. The iron ore combines with hot carbon monoxide gas.

iron ore + carbon monoxide → iron + carbon dioxide

$$Fe_2O_3(s) + 3CO(g) \rightarrow 2Fe(s) + 3CO_2(g)$$

The iron ore loses oxygen, so it is reduced. The carbon monoxide gains oxygen, so it is oxidised. Extracting metal from ore is a redox reaction.

Fig. 36.4   A battery of cells used to decompose sodium chloride. The cells contain sodium chloride solution. An electric current is passed through the solution. Sodium and chlorine are produced.

RATE OF REACTION

Chemical reactions happen at different rates. You can measure the rate of a reaction by the rate of disappearance of the reactants, or by the rate of appearance of the products.

Reactions go at different rates

By now, you will probably have come across a number of different chemical reactions. A chemical reaction is a process by which one or more substances called **reactants** changes into one or more other substances, called **products**.

Some examples of chemical reactions which you may have met are:

A magnesium + sulphuric acid → magnesium sulphate + hydrogen

B iron + oxygen → iron oxide

C silver nitrate + hydrochloric acid → silver chloride + nitric acid

D copper oxide + sulphuric acid → copper sulphate + water

You may have noticed that these reactions occur at different speeds. C, for example, occurs instantaneously. A is a rapid reaction. D is slow. B is extremely slow, especially if the iron is dry. We say that these reactions have different **rates of reaction**.

INVESTIGATION 37.1

The thiosulphate cross

If you mix sodium thiosulphate solution with hydrochloric acid, a reaction occurs. One product is the element sulphur. This makes the mixture cloudy.

1 Copy the table.

Experiment	1	2	3	4	5	6	7
Volume of sodium thiosulphate (cm³)	40	30	20	40	40	40	40
Volume of water (cm³)	0	10	20	5	7.5	0	0
Volume of acid (cm³)	10	10	10	5	2.5	10	10
Temperature of sodium thiosulphate (°C)	room	room	room	room	room	40	60
Time (s)							

2 Draw a cross, with a dark pencil, on a piece of white paper. Stand a 100 cm³ beaker on the cross.
3 Put 40 cm³ of sodium thiosulphate solution into the beaker. Add 10 cm³ of hydrochloric acid, and start timing. Stop timing when the mixture is so cloudy that you can no longer see the cross. Fill in this result in the table.
4 Rinse the beaker very thoroughly. Perform experiments 2 to 7, rinsing the beaker after each experiment. Record each of your results.

Rate can be measured

Reaction A, above, is quite a rapid reaction. You can tell this because the magnesium disappears quickly. Reaction D is slow, because the copper oxide disappears slowly. Reaction B is very slow. The patches of iron oxide take days to form. Reaction C is instantaneous, because the silver chloride forms immediately when the solutions mix.

So we use two different ways to judge how fast a reaction takes place. Rate of reaction is **the rate of disappearance of a reactant or the rate of appearance of a product**.

Fig. 37.1 As hydrochloric acid is added to sodium thiosulphate solution, a cloudy precipitate of sulphur is formed which obscures the cross. You can measure the speed of the reaction by timing how long it takes for the cross to disappear.

Questions

1 What can you conclude by comparing results for experiments 1, 2 and 3?
2 What can you conclude by comparing results for experiments 1, 4 and 5?
3 What can you conclude by comparing results for experiments 1, 6 and 7?
4 Which are you timing – the disappearance of a reactant, or the appearance of a product?
5 Suggest how you would expect your results to change if you did the experiment in a 250 cm³ beaker, but used the same quantities of reactants.

Factors affecting the rate of reaction

Hydrochloric acid reacts with marble chips (calcium carbonate). One product is carbon dioxide gas. You can measure the rate of reaction by timing how long it takes to make 10 cm³ of gas. You are going to try altering several factors, to find out which ones affect the rate of the reaction.

1 Copy the table below.

Experiment	1	2	3	4	5	6	7
Vol. acid (cm³)	50	50	50	50	100	50	50
Temperature of acid (°C)	room temp	40	60	room temp	room temp	room temp	room temp
Concentration of acid (mol/dm³)	2	2	2	2	2	1	0.5
Marble chip	whole	whole	whole	crushed	whole	whole	whole
Time for 10 cm³ of CO_2 to collect							

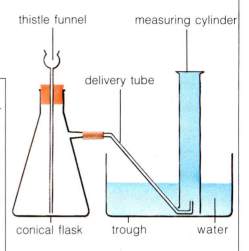

Fig. 37.2 Apparatus for generating carbon dioxide. The thistle funnel is optional. If you do not have one, use a solid bung in the conical flask. Take out the bung to put in the marble chips and acid, then replace it quickly.

thistle funnel measuring cylinder

delivery tube

conical flask   trough   water

2 Set up the apparatus as in the diagram. The inverted measuring cylinder full of water is to collect the gas. Move it away from the outlet of the delivery tube to begin with.

3 Remove the bung. Put 50 cm³ of acid and one marble chip into the flask. Replace the bung. 5 s later, place the inverted measuring cylinder over the mouth of the delivery tube. Time how long it takes collect 10 cm³ of gas.
This is experiment 1. Fill in your result.

4 Perform experiments 2 to 7. Fill in your results for each one. In each case, choose marble chips of similar size.

For experiments 2 and 3, *carefully* warm the acid in a beaker before adding it to the marble chip.

For experiment 4, crush one marble chip in a mortar.

For experiment 6, mix 25 cm³ of 2 mol/dm³ acid with 25 cm³ of water. This makes 1 mol/dm³ acid.

For experiment 7, mix 12.5 cm³ of 2 mol/dm³ acid with 37.5 cm³ of water. This makes 0.5 mol/dm³ acid.

Questions

1 Which factor, or factors, increased the rate of reaction?

2 Which factor, or factors, decreased the rate of reaction?

3 Which factor, or factors, did not affect the rate of reaction?

4 Why were marble chips of similar size used?

5 Why was the measuring cylinder not over the delivery tube while the bung was replaced?

6 Which of these conditions would you choose to make the reaction go as fast as possible?
Acid temperature: room temp.   40 °C   60 °C
Acid concentration: 2 mol/dm³   1 mol/dm³   0.5 mol/dm³
Marble chip: whole   crushed   whole

7 Which conditions would you choose, from those in question 6, to make the reaction go as slowly as possible?

8 Write a word equation and then a balanced molecular equation, for the reaction involved.

Rate of reaction matters to manufacturers

A slow reaction can be irritating to wait for in a laboratory. But in industry it is much more serious. A slow reaction means slow production. This can mean less profit. Chemical engineers have to understand the factors affecting rate of reaction and use them to make the product form more quickly.

Questions

1 Rank these reactions in order of increasing rate:
a sodium hydroxide solution neutralising hydrochloric acid
b wood rotting
c milk turning sour
d an egg cooking in boiling water
2 Zinc and sulphuric acid react to produce zinc sulphate and hydrogen.

a Write a word equation for this reaction.
b Write a balanced molecular equation for this reaction.
c What are the reactants?
d What are the products?
e List four ways in which you could judge the rate of this reaction.
f How could you speed up this reaction?
g How could you slow down this reaction?

38 CATALYSTS

Catalysts are substances which speed up chemical reactions without being changed themselves.

Catalysts speed up reactions without being changed themselves

One way of making oxygen in the laboratory is by the decomposition of hydrogen peroxide, H_2O_2.

hydrogen peroxide $\longrightarrow$ water + oxygen

$$2H_2O_2(aq) \longrightarrow 2H_2O(l) + O_2(g)$$

Under normal conditions, this reaction is very slow. It can be speeded up by using more concentrated hydrogen peroxide, or by heating it. But a much easier way of speeding it up is to add **manganese (IV) oxide**, MnO_2. Manganese (IV) oxide is a fine, black powder. If you add it to hydrogen peroxide solution – even if it is cold and dilute – the hydrogen peroxide starts to decompose rapidly. Surprisingly, the manganese (IV) oxide is not used up. When the reaction finishes, the black powder is still there.

Manganese (IV) oxide is an example of a **catalyst**. A catalyst is **a substance which speeds up a chemical reaction without being changed or used up.**

Catalysts are important in industry

Catalysts are very important to chemical manufacturers. They allow more product to be made in a given length of time. They are not used up, so they do not have to be continually added to the reactants. But they do have to be cleaned every now and then. This is because the activity of a catalyst involves its surface. Dirt or impurities on a catalyst's surface will stop it acting as efficiently as it should. Industry uses many different catalysts to speed up the reactions it relies on.

Fig. 38.2 Biological washing powders contain enzymes which break down stains such as coffee, egg, blood or sweat.

Biological catalysts are called enzymes

Living organisms have hundreds of different chemical reactions going on inside them. Some of these reactions need to happen quickly. The reactions could be speeded up by heating the reactants. But living cells are rapidly killed if their temperature goes too high.

Fig. 38.1 The conical flask contains hydrogen peroxide solution. Without a catalyst nothing happens. But when a small pinch of manganese (IV) oxide is added, the hydrogen peroxide rapidly decomposes to water and oxygen. The manganese (IV) oxide is not used up in this reaction.

So living organisms use catalysts to speed up their reactions. These catalysts are called **enzymes**. Each different reaction happening in a living organism has a different enzyme catalysing it. So there are hundreds of different kinds of enzyme.

Enzymes are proteins and consist of very large molecules. These molecules have special shapes which are important to their ability to act as catalysts. If enzymes are heated above 45°C, their molecules lose their shapes, and so they don't work well as catalysts at higher temperatures.

Catalysing the decomposition of hydrogen peroxide with an enzyme

You have seen how hydrogen peroxide can be made to decompose quickly by using manganese (IV) oxide as a catalyst. There is also a biological catalyst which will speed up this reaction. It is called **catalase**. Catalase is found in many different kinds of cell. You can find it in apples, yeast, meat (muscle cells) – almost anything! One of the places where there is an especially large amount of catalase is in liver.

Catalase is needed in living cells because hydrogen peroxide is poisonous. Hydrogen peroxide is thought to be made as a by-product of several reactions in cells. It must be decomposed instantly, before it can do any damage. This is what catalase does.

1 Read through this experiment and then design and draw up a results chart.
2 Measure out 20 cm³ of hydrogen peroxide into each of four boiling tubes.
3 Cut four small pieces of fresh liver. They must all be of the same size, around 0.5 cm².
4 Cook two of the pieces of liver in boiling water for about 5 min. Remove them and cool.
5 Grind one raw and one cooked piece of liver to a paste. Keep them separate from one another.

You now have four samples of liver – whole raw, ground raw, whole cooked and ground cooked.

Make sure that you are thoroughly organised before you go any further – things will happen quickly!

6 Start a stopclock. Put your four samples of liver into the four tubes of hydrogen peroxide solution. Try to do them all at the same time. After 2 min, measure the height of the froth in the boiling tube, and record it in your results chart.
7 Light and then blow out a splint.

Push the glowing splint through the froth in one of the tubes, and record what happens.

Questions

1 Write a word equation, and a balanced molecular equation, for the reaction which happened in your boiling tubes.
2 The liver contained a catalyst. What is the name of this catalyst? What type of catalyst is it?
3 Why does liver contain this catalyst?
4 Which type of liver made the reaction happen fastest? Explain why you think this is.
5 Which type of liver was least effective in speeding up the reaction? Explain why you think this is.

Questions

1 This question is about the decomposition of hydrogen peroxide to water and oxygen.
a Give two ways in which the decomposition of hydrogen peroxide could be speeded up.
b Give three ways in which the decomposition of hydrogen peroxide could be slowed down.
c Manganese (IV) oxide acts as a catalyst in this reaction. How could you prove that the manganese (IV) oxide is not used up in the reaction flask?
d Is manganese (IV) oxide
  i a reactant?
  ii a product in this reaction?
e Do you think that manganese (IV) oxide would be a better catalyst in lump form, or in powder form? Explain your answer.
2 New cars must by law have catalysts fitted into their exhaust systems.
a Why do you think that this is done?
b Find out why using leaded petrol prevents such catalysts being used.

3 Enzymes are biological catalysts. All enzymes are proteins. Protein molecules are easily destroyed by high temperatures.
The graph below shows how a particular biological reaction is affected by temperature. The reaction is catalysed by an enzyme.
  a At which temperatures did the reaction take place most slowly?
  b At which temperature did the reaction take place most quickly?
c What happened to the reaction as the temperature was raised from 10 °C to 20 °C?
d What do you think began to happen to the enzyme at 45 °C?
e Explain why the reaction got slower as the temperature was raised from 45 °C to 60 °C.
f Human body temperature is normally around 37 °C. If you are very ill, your temperature can go up to 40 °C. Why is this dangerous?

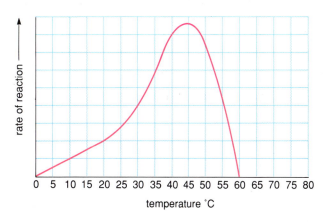

temperature °C

39 SPEEDING UP REACTIONS

Knowing which factors speed up reactions, and why, can be used to make industrial processes more efficient.

Why do some factors affect rate of reaction?

The marble and hydrochloric acid reaction shows you that, when a solid reacts with a solution, the reaction goes faster if:

• the solid is in smaller particles
• the solution is hotter
• the solution is more concentrated.

To understand why this is, you must think about the particles present.

The marble is solid. Its particles are vibrating slightly, but cannot move around freely. The hydrochloric acid is made up of several different sorts of particles. It contains water molecules, H^+ ions and Cl^- ions. These are all in constant motion. They are moving around freely.

For the reaction to occur, the H^+ ions must collide with the marble particles. The harder they hit, the more likely it is that the particles will react together. Anything which makes the acid and marble particles collide more often, or collide more violently, will speed up the reaction.

Finer powdered marble has a larger surface area. So acid particles collide with it *more often.*

Hotter acid has faster moving particles. They therefore collide with the marble *more often* and *more violently*. Only particles that possess a certain minimum amount of energy (called the **activation energy** – see Topic 34) will react when they collide. Heating the acid means that more particles have enough energy to react when they collide, so the reaction happens faster.

More concentrated acid has more acid particles per cubic centimetre. So acid particles will collide with the marble *more often.*

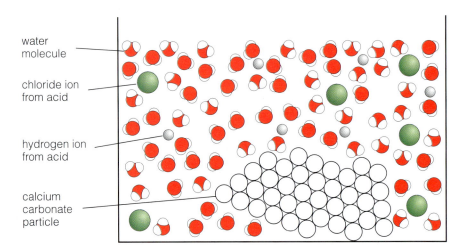

Fig. 39.1 In a lump of marble (calcium carbonate) only the particles on the surface can come into contact with the acid. So only they can react. If you break the marble apart, more of it can react.

water molecule
chloride ion from acid
hydrogen ion from acid
calcium carbonate particle

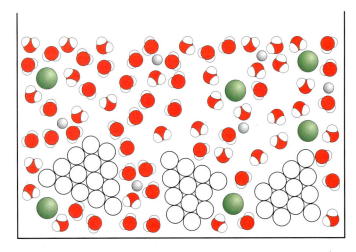

Fig. 39.2 If the lump of marble is broken up into smaller bits, more of its particles can come into contact with the acid, so the reaction will proceed more quickly.

Some reactions are between gases. An example is the reaction between ammonia and oxygen on the next page. How can a gas be made more concentrated? 'More concentrated' means with more particles in a certain volume. The way to force more gas particles into a certain volume is to **pressurise** the gas. Putting gases under greater pressure makes the particles more concentrated, so they collide with each other more often and therefore react faster.

Living organisms speed up their reactions by keeping warm

The reactions inside living organisms are called **metabolic reactions**. The speed at which these reactions take place is known as the **metabolic rate**. The metabolic rate of an organism determines how active it can be. If the metabolic rate is very slow, then the organism is also slow and sluggish. A faster metabolic rate will allow the organism to move more swiftly and be more alert.

One way in which organisms speed up their metabolic rate is by using enzymes as catalysts. In fact, without enzymes, most metabolic reactions would go so slowly that they would never happen at all! All living things rely on enzymes to make their metabolic reactions happen.

Another way of speeding up meta-bolic rate is to keep warm. Most living things are the same temperature as their surroundings. They are called **poikilothermic** or cold-blooded organisms. When the weather is cold, their metabolism is very slow. Poikilo-thermic animals, such as insects, are very slow and sluggish in winter. But in warm weather, their reactions happen much faster. They can fly actively.

Two groups of animals go one better. They can keep their temperature warm all the time. These are the birds and mammals. They are **homeothermic** or warm-blooded animals. Mammals keep their temperature around 37°C, even in cold weather. They can stay active all the year round. They use the chemical energy in food to produce the heat energy needed to stay warm.

Fig. 39.3   A smooth newt is poikilothermic whereas a red deer is homeothermic.

Making nitric acid

Ammonia is a very important raw material for many chemical processes. One especially important one is the manufacture of **nitric acid**.

This process involves three stages.

1   Ammonia reacts with oxygen from the air to make nitrogen monoxide.

ammonia + oxygen $\longrightarrow$ nitrogen monoxide + water

$4NH_3(g) + 5O_2(g) \longrightarrow 4NO(g) + 6H_2O(g)$

2   Nitrogen monoxide reacts with more oxygen from the air to give nitrogen dioxide.

$2NO(g) + O_2(g) \longrightarrow 2NO_2(g)$

3   Nitrogen dioxide reacts with water to make nitric acid.

$3NO_2(g) + H_2O(l) \longrightarrow 2HNO_3(aq) + NO(g)$

Fortunately, reactions **2** and **3** are extremely rapid. They need no help from people controlling the conditions. But reaction **1** is very slow. Conditions must be produced in which the reaction will take place more rapidly. This involves:

a increasing pressure to around seven times atmospheric pressure

b raising the temperature to between 800 and 1000°C

c using a catalyst. This is gauze woven from strands of the precious metals platinum and rhodium.

It is a good job that catalysts are not used up! Each of these pieces of gauze costs around half a million pounds!

The production of nitric acid by this method is a continuous process. The reactants flow into the reactor vessel, and the products flow out continually. This goes on for several weeks or months. The reactor vessel is then shut down and cleaned. But the remains from cleaning are not thrown away. They contain enough tiny fragments of catalyst to make it worthwhile to reclaim them.

Questions

1 a Explain why the reaction between ammonia and oxygen is speeded up by:
  i increased pressure
  ii increased temperature.
  b In this reaction, platinum–rhodium gauzes are used as catalysts.
  i State two things about the gauze which justifies its being called a catalyst.
  ii Why are the catalyst metals made into a gauze, and not used in lump form?
  iii Why do the gauzes need cleaning periodically?

Fig. 39.4   A platinum–rhodium gauze used as a catalyst in the commercial production of nitric acid.

40 THE HABER PROCESS

Ammonia is made from nitrogen and hydrogen by the Haber process. This process is reversible.

Nitrogen and hydrogen react very slowly under normal conditions

In the preceding Topic you have seen how nitric acid is made using ammonia as its main starting material. How is ammonia made? The formula of ammonia is NH₃, so the obvious solution seems to be to react nitrogen and hydrogen together.

$$N_2 + 3H_2 \longrightarrow 2NH_3$$

This sort of reaction is called a **synthesis**. Both the reactants are readily available. The nitrogen can be obtained from the air and the hydrogen is obtained from oil products by a process called **cracking**, or from natural gas.

Unfortunately the reaction between nitrogen and hydrogen is extremely slow under normal conditions, because the triple bond holding the two nitrogen atoms together in an N_2 molecule is very strong. The reaction is slow because it has a high **activation energy**. The problem of how to speed up the reaction, so that a reasonable amount of ammonia is produced in a reasonable time, was solved in Germany by Fritz Haber and Karl Bosch in 1908 and 1909, and was developed into a full scale industrial process in the next few years. You should by now be able to guess the conditions that are used in order to increase the rate of the reaction. Close this book and try to write down the conditions you would choose, and your reasons for choosing them, before you read on.

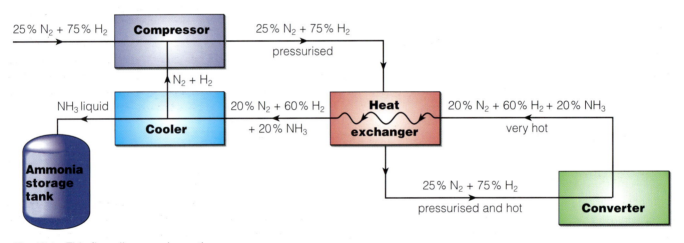

Fig 40.1   This flow diagram shows the steps in the manufacture of ammonia.

Increasing pressure can speed up gas reactions

The reaction between nitrogen and hydrogen is used to make ammonia. Ammonia is an important material in the fertiliser industry.

nitrogen + hydrogen ⟶ ammonia

$$N_2(g) + 3H_2(g) \longrightarrow 2NH_3(g)$$

This is called the **Haber process**.

You know four ways to speed up a reaction:

1 *Use a catalyst* Iron is used as a catalyst in the Haber process. The reactor vessel is filled with short lengths of iron tube.
2 *Use solids in powder form* But in this case both reactants are gases.

3 *Increase the temperature* The gases are heated to 400°–500°C.
4 *Increase the concentration of the reactants* How do you increase the concentration of a gas? The answer is to **pressurise** it. High pressure forces the gas particles closer

together. This makes the gas more concentrated. In the Haber process, the pressure used is around 200 times atmospheric pressure. This is very costly, but it is well worth it in terms of increased production of ammonia.

Fig. 40.2   Part of ICI's Number 4 ammonia plant at Billingham. The shortest of the three grey columns at the left is the ammonia converter, where hydrogen and nitrogen combine together to form ammonia. The reaction takes place at 200 times atmospheric pressure. The reaction is exothermic, and the heat from it is used to heat fresh supplies of hydrogen and nitrogen, to speed up their rate of reaction. This happens in the heat exchangers, fed by the yellow pipes at the bottom right of the picture.

The Haber process makes ammonia at an economical rate

The following conditions are used:

1 The nitrogen and hydrogen gases are mixed in the ratio 25% nitrogen to 75% hydrogen. This mixture is then **pressurised** in the **compressor**. This increases the **concentration** of the gases. The pressure used is usually 200 times atmospheric pressure.

2 The nitrogen and hydrogen gases are **heated**. This is done in the **heat exchanger**. The gases are heated to 450°C.

3 A **catalyst** is used. The catalyst is iron metal in the form of short lengths of iron tube. The catalyst is contained in a vessel called the converter. Some of the nitrogen and hydrogen reacts together on the surface of the catalyst in the converter vessel while some remains unreacted. The mixture leaving the converter vessel contains about 20% ammonia, 20% unreacted nitrogen and 60% unreacted hydrogen.

The reaction $N_2 + 3H_2 \longrightarrow 2NH_3$ is **exothermic**. This means that the mixture of gases leaving the converter vessel is hotter than the mixture of gases that went in. This extra heat is used to heat up the gases that are going into the converter vessel. This takes place in the heat exchanger. The next operation is to remove the ammonia from the mixture, which is done by cooling the mixture so the ammonia becomes a liquid. It can then be piped off to storage tanks while the unreacted nitrogen and hydrogen are sent round again giving them another chance to form ammonia. Make sure you can follow all this on Figure 40.1.

The synthesis of ammonia is reversible

20% of ammonia doesn't seem to be a great return for such efforts. The unreacted nitrogen and hydrogen do go round again but perhaps this would be unnecessary if the reaction were made still faster? Unfortunately the answer is no. Some reactions are **reversible** and this is one of them. As soon as ammonia begins to form in the reactor vessel by the reaction

$$N_2 + 3H_2 \longrightarrow 2NH_3$$

some of the ammonia formed begins to break down by the reaction

$$2NH_3 \longrightarrow N_2 + 3H_2$$

So the best way to write the equation is like this

$$N_2 + 3H_2 \rightleftharpoons 2NH_3$$

The double arrow $\rightleftharpoons$ means that the reaction is reversible. No matter what you do you can't get 100% conversion of reactants to products. The reaction always reaches a balanced position where reactants are being converted to products and products are breaking down to reactants at the same rate. This position is called **equilibrium.**

A further factor to be considered is that the proportion of reactants and products at equilibrium is not fixed. It depends on the conditions. In the explanation that follows we will call $N_2 + 3H_2 \longrightarrow 2NH_3$ the **forward reaction** and we will call $2NH_3 \longrightarrow N_2 + 3H_2$ the **backward reaction**.

At higher temperatures the rate increases but less ammonia forms

The forward reaction is exothermic, so the backward reaction is endothermic. For *any* reversible reaction, higher temperatures favour the endothermic reaction. Therefore in the Haber process, the higher the temperature, the more the backward reaction is favoured and the less ammonia there will be in the mixture when equilibrium is reached. 450°C is a compromise temperature – if a higher temperature were used the *amount* of ammonia formed would be too low, but if a lower temperature were used the *rate* of formation of ammonia would be too slow.

At higher pressures the rate increases and more ammonia forms

If a reversible reaction involves one or more gases, the number of molecules involved is important. High pressure favours the side of the equation where there is the **smaller** number of gas molecules. In the Haber process, three hydrogen molecules react with one nitrogen molecule (a total of four reactant molecules) to form two ammonia molecules, so high pressure causes there to be more ammonia in the mixture when equilibrium is reached. High pressure therefore causes more ammonia to be formed *and* increases the *rate* at which it is formed.

The presence of a catalyst doesn't affect the amounts of reactants and products in the mixture when equilibrium is reached, but it does affect how **quickly** equilibrium is reached. The catalyst affects the rate only.

Ammonia is mainly used to make fertilisers

As we have seen, ammonia is used to make nitric acid. If nitric acid is reacted with ammonia, a **neutralisation** reaction takes place and a salt called **ammonium nitrate** is formed.

$$NH_3 + HNO_3 \longrightarrow NH_4NO_3$$

Questions

1 a What are the reactants in the Haber process and how are they obtained?

b What is the product in the Haber process and what is it used for?

c What catalyst is used? Why is a catalyst necessary?

d Why is a temperature as high as 450°C used?

e Why are even higher temperatures not used?

f Give two advantages in running the Haber process at high pressure.

2 What do the terms 'reversible reaction' and 'equilibrium' mean?

The rate of a reaction can be studied by measuring the amount of product formed every minute, or by measuring the amount of reactants left every minute.

Measuring the rate of reaction of nitric acid with copper carbonate

Often, rates of reaction are studied by measuring the amount of product formed each minute, until the reaction is finished. This method could be used to study the reaction between copper carbonate and nitric acid. This reaction produces carbon dioxide.

This reaction could be carried out by using nitric acid of varying concentration and temperature. An excess of copper carbonate would be used to make sure that all of the acid reacts. The sort of results you might obtain are shown in the table.

Look particularly at the following points:

a Note that the rate of carbon dioxide formation slows down in all the reactions. This is because, as the nitric acid is used up, it becomes less concentrated. This slows the reaction down.

b Compare experiments 1 and 2. They both have the same *concentration* and *temperature* of acid. So they both start at the same rate. But experiment 2 has twice the *amount* of acid. So this reaction carries on for longer, and produces twice the amount of carbon dioxide.

c Compare experiments 1 and 3. Both have the same *amount* of acid, so both produce the same amount of carbon dioxide. But experiment 1 has *more concentrated* acid, so the carbon dioxide forms more quickly.

d Compare experiments 1 and 4. Both have the same *amount* of acid, so both produce the same amount of carbon dioxide. But experiment 4 has *hotter* acid, so the carbon dioxide forms more quickly.

copper carbonate + nitric acid → copper nitrate + water + carbon dioxide

$$CuCO_3(s) + 2HNO_3(aq) \rightarrow Cu(NO_3)_2(aq) + H_2O(l) + CO_2(g)$$

Experiment	1	2	3	4
Volume of acid (cm³)	50	100	50	50
Volume of water (cm³)	0	0	50	0
Temperature (°C)	20	20	20	60
Amount of CO₂ produced (cm³) at:				
1 min	60	60	30	95
2 min	85	115	56	99
3 min	95	155	77	100
4 min	98	180	90	100
5 min	99	192	96	100
6 min	100	196	98	100
7 min	100	198	99	100
8 min	100	200	100	100
9 min	100	200	100	100

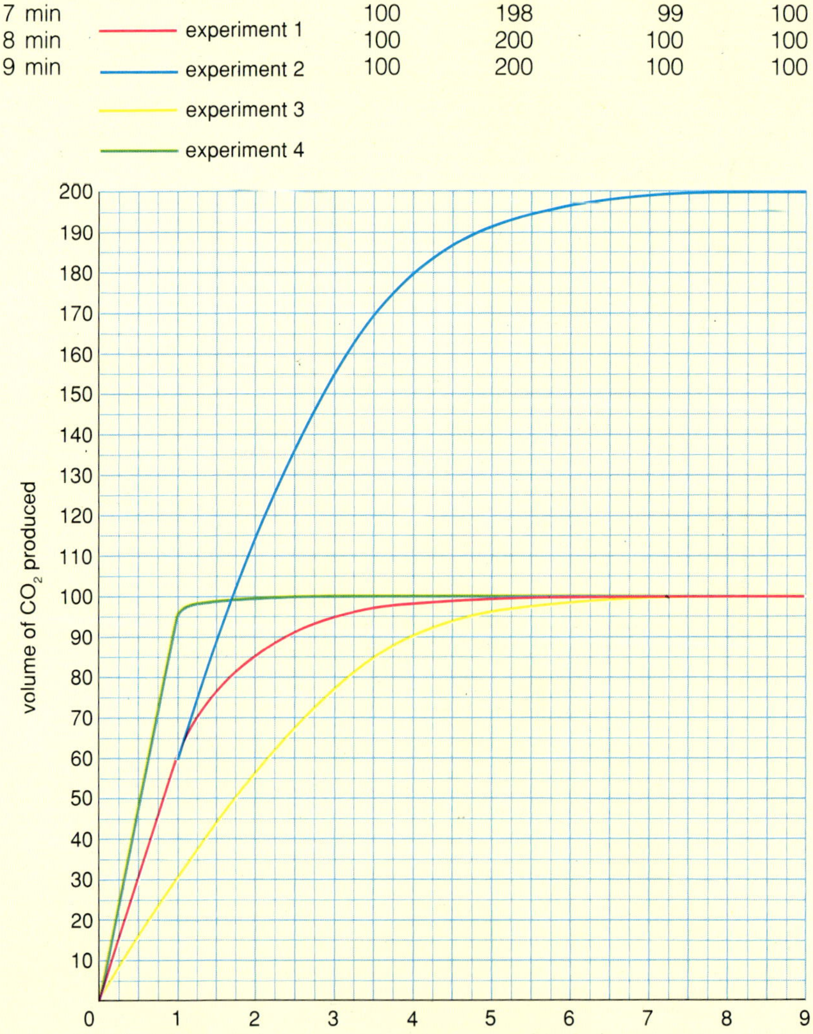

—— experiment 1
—— experiment 2
—— experiment 3
—— experiment 4

Fig. 41.1

The reaction between magnesium and sulphuric acid

To follow this reaction, you are going to measure the rate of evolution of hydrogen gas. The equation for the reaction is:

magnesium + sulphuric acid ⟶ magnesium sulphate + hydrogen

$$Mg(s) + H_2SO_4(aq) \longrightarrow MgSO_4(aq) + H_2(g)$$

This experiment can be performed with syringe apparatus. You may be shown this. But in the method described here you will collect the hydrogen in a measuring cylinder.

Fig. 41.2 Generating hydrogen

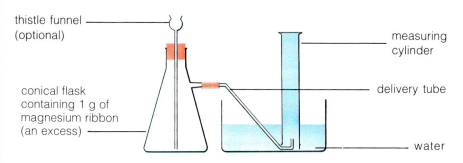

1 Assemble the apparatus as shown in the diagram.
2 Copy the chart. It should continue for 10 min.

Experiment	1	2	3
Volume of acid (cm³)	25	50	25
Volume of water (cm³)	25	0	25
Temperature of liquid (°C)	room	room	50
Volume of gas collected at: 1 min			
2 min			

3 Move the measuring cylinder away from the delivery tube.
4 Remove the bung from the flask. Add 25 cm³ of acid and 25 cm³ of water. Replace the bung. After 2 s, put the measuring cylinder back over the delivery tube. When the first bubble reaches the top of the measuring cylinder, start timing.
5 Record the volume of gas in the measuring cylinder every minute for 10 min.
6 Perform experiments 2 and 3 and fill in the results.
7 Plot these results as a line graph. Put time on the horizontal axis, and volume of gas collected on the vertical axis. Draw all three curves on the same graph.

Questions

1 Why is excess magnesium used?
2 Why does the reaction slow down in each case?
3 Why will the reaction eventually stop?·
4 Experiment 2 probably produced hydrogen more rapidly than experiment 1. It will eventually produce twice as much gas in total. Why is this?
5 Experiment 3 probably produced hydrogen more rapidly than experiment 1. But eventually both experiments will produce the same amount of gas. Why is this?
6 Imagine that the conditions of experiment 1 were repeated. However, this time, 1 g of finely powdered magnesium was added. Draw a fourth line on your graph to show the results you would expect. Explain your choice.

The digestion of starch by amylase

Amylase is an enzyme that catalyses the reaction in which starch breaks down to a sugar called maltose. This happens in your mouth, because saliva contains amylase.

1 Label three boiling tubes A, B and C. Measure 10 cm³ of starch solution into each.
2 Stand tube A in a rack in a water bath at 35 °C. Stand tube B in a rack on the bench. Stand tube C in a rack in a refrigerator. Measure and record the temperature on the bench and in the refrigerator. Leave all the tubes for at least 10 min.
3 While you are waiting, put a drop of iodine solution into each space on a spotting tile. Label the rows on the tile A, B and C. Design and draw up a results chart.
4 When the tubes have been left for 10 min, put 10 cm³ of amylase solution into each one. Mix the starch and amylase thoroughly. Try to do this at the same time for all three, and start a stop clock. Put the tubes back into their appropriate places.
5 Every 2 min, take a small sample from each tube, and add it to one of the iodine solution drops on the tile. Record your results as you go along. Keep doing this until at least two of the tubes no longer contain any starch.

Questions

1 Is starch a reactant or a product in this reaction?
2 In which tube did the starch disappear most quickly? Explain why you think this is.
3 What product would you expect to be formed in this tube?
4 In which tube did the starch disappear most slowly? Explain why you think this is.
5 If you had kept one of the tubes at 90 °C, would you expect the starch to disappear quickly, slowly, or not at all? Explain your answer.
6 If you had added water instead of amylase solution to one tube, would you expect the starch to disappear quickly, slowly, or not at all? Explain your answer.

Questions

1 Explain each of the following observations:
   a Hot substances react faster than cold ones.
   b Powdered zinc reacts faster than zinc granules with acid.
   c Zinc reacts faster with fairly concentrated acids than with dilute acids.
   d Zinc reacts faster with strong acids than with weak acids.
   e The chemical industry is constantly searching for new catalysts.
   f Pressuring gaseous reactants affects rate in the same way as concentrating dissolved reactants.
   g The Haber process has been run at pressures of about 1000 atmospheres. However this greatly increases capital expenditure and operating costs compared with running the process at atmospheric pressure.

2 The digestion of starch is a chemical reaction in which large starch molecules are broken down to smaller ones. The equation for the reaction is:

$$\text{starch} + \text{water} \rightarrow \text{maltose}$$

This reaction occurs in your mouth. Saliva contains an enzyme called amylase which catalyses this reaction.
   a What is the temperature inside a person's mouth?
   b What would happen to the rate of this reaction if the temperature was much lower than this? Explain your answer.
   c What would happen to the rate of this reaction if the temperature was much higher than this? Explain your answer.
   d Saliva contains the enzyme amylase, which catalyses this reaction. What other important substance is present in saliva which will help the reaction to proceed?
   e Teeth can also help this reaction to proceed faster. Explain how they can do this.

EXTENSION

3 The graphs show the results of five experiments involving the reaction between hydrochloric acid and excess marble. Details of the experiments are given in the chart beneath the graphs.

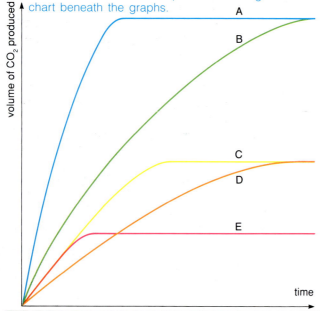

Experiment	1	2	3	4	5
Vol. of acid (cm³)	60	30	30	15	60
Vol. of water (cm³)	60	30	30	15	60
Temp. of liquids (°C)	40	20	40	40	40
Marble (w = whole) (p = powdered)	w	w	w	w	p

   a For each experiment (1–5) identify the correct graph A–E. Give reasons for each of your choices.
   b Which factor is the same in every experiment?
   c Which reaction finished first?

4 An experiment was done to find out what factors affect the rate of decomposition of hydrogen peroxide solution. The pupil had a solution of hydrogen peroxide, distilled water, manganese (IV) oxide powder, some measuring cylinders, a gas generator and a means of collecting gas. She put various mixtures into her gas generator, and measured how much gas was produced each minute for 10 min. The results are shown below.

	Volume of gas collected (cm³)			
Experiment	1	2	3	4
Time (min)				
1	40	80	70	0
2	60	120	77	0
3	70	140	79	0
4	75	150	80	0
5	78	155	80	0
6	79	158	80	0
7	79	159	80	0
8	80	159	80	0
9	80	160	80	0
10	80	160	80	0

In experiment 1, the pupil used 50 cm³ of hydrogen peroxide solution, 50 cm³ of water (both at 20°C) and a pinch of manganese (IV) oxide. Unfortunately, she forgot to note down what she put into experiments 2, 3 and 4. The total volume of liquid in the flasks was always always 100 cm³.
   a What was the gas collected each time?
   b Write an equation for the reaction.
   c What part does manganese (IV) oxide play in this reaction?
   d Draw a diagram of the equipment which might have been used.
   e What might have gone into the flask in experiment 2? Explain your answer.
   f What might have gone into the flask in experiment 3? Explain your answer.
   g What might have caused the result in experiment 4?

THE EARTH

42 THE STRUCTURE OF THE EARTH

The Earth is made up of several different layers. Evidence for this comes from the behaviour of earthquake waves.

The Earth's crust is like a thin skin

We think of the Earth under our feet as being a solid, unchanging structure. But this is not quite true. The Earth is not completely solid and it changes a great deal.

The outer layer of the Earth, on which we live, is called the **crust.** The crust is much thicker under the continents than under the oceans. Continental crust is between 25 and 40 km deep. Oceanic crust is between 5 and 10 km deep. These distances may sound quite large, but they are tiny compared with the size of the Earth. The radius of the Earth is about 6378 km, so the crust is just a very thin skin on the surface.

The thin oceanic crust is much denser than the thick continental crust.

The mantle makes up most of the Earth

Beneath the crust, there is a very thick layer called the **mantle**. The mantle is about 2800 km thick. It is made up of rock containing a lot of iron. Together, the upper part of the mantle and the crust, are sometimes called the **lithosphere**.

The rocks in the mantle are very dense, because of the huge crushing forces of all the rocks above. They are also hot. Although the outer part of the mantle is fairly solid, the great forces can make it flow slowly. The rocks of the inner layers of the mantle are very hot, and they are partly liquid. This molten rock is called **magma**. Sometimes, this molten rock escapes through cracks in the crust. It is then called **lava**.

The core is partly liquid

The centre of the Earth is called the **core**. The core is about 6950 km in diameter. The inner part is solid, but the outer core is liquid. The core is mostly made up of iron and nickel. It is thought that the movements of the outer, liquid, core may be responsible for producing the Earth's magnetic field.

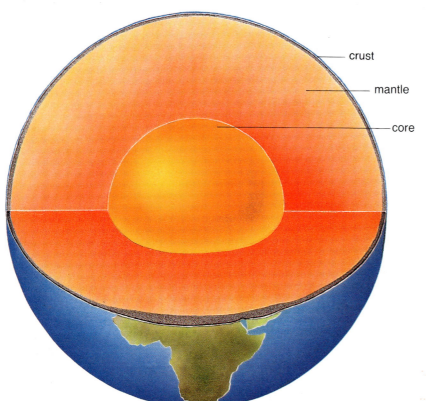

Fig. 42.1 The structure of the Earth

crust

mantle

core

> ### DID YOU KNOW?
>
> Although the volume of the inner core is only 16% of the Earth's total volume, its mass makes up nearly 33% of the Earth's total mass. This is because the materials from which it is made are very dense.

Fig. 42.2 Continental crust is a thick layer of relatively light rock, floating on the mantle. Oceanic crust is denser, and forms a thinner layer. The boundary between the crust and the mantle is called the Mohorovic discontinuity (Moho for short) after the Yugoslav geologist who discovered it.

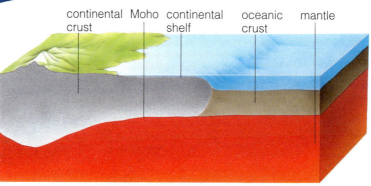

continental crust   Moho   continental shelf   oceanic crust   mantle

The behaviour of earthquake waves gives clues about the structure of the Earth

No-one has been able to take samples from the inside of the Earth. Even the very deepest holes which have ever been made by humans only go into the crust. Everything we know about the inner layers of the Earth has been worked out indirectly.

Much of the evidence which suggests that the Earth is made up of layers comes from the study of **earthquake waves**. Earthquakes produce longitudinal waves and transverse waves. The longitudinal waves are called **P waves**. They are like sound waves. They produce alternating areas of compression and rarefaction in the material through which they travel. P waves travel very quickly. They are the fastest waves produced by an earthquake. This is why they are called P waves. The P stands for primary, because they are the first waves to be detected.

The transverse waves are called **S waves**. They travel like a wave in a rope which is shaken up and down. S waves travel more slowly than P waves. S stands for secondary, because these waves are not detected until after the P waves.

When an earthquake happens, P waves and S waves travel through the Earth. The waves are detected by **seismographs**. Seismographs pick up the P waves first, then the S waves. By comparing the time taken for the two different kinds of waves to reach a seismograph, scientists can work out where the earthquake happened.

But the P and S waves also give information about the inside of the Earth. For example, it is found that the waves suddenly begin to travel faster once they are about 25 to 40 km below the surface. This suggests that the density and composition of the rocks suddenly changes at this level. This change is called the **Mohorovic discontinuity**. It is at the boundary between the crust and the mantle.

Another change in the behaviour of the waves happens much deeper in the Earth. S waves cannot travel through the centre of the

Earth at all. This suggests that it is at least partly liquid. So it is believed that there is a liquid outer core. P waves can travel through this region. The speed at which they travel suggests that the outer core is mostly made of iron.

Further evidence for the existence of a liquid, iron-containing outer core comes from studies of the Earth's magnetic field. In theory, the magnetic field could be produced by the movement of huge quantities of molten iron at this level in the Earth. What causes this movement is still a mystery. It is probably a combination of convection currents, caused by uneven heating of the Earth's interior, and forces set up by the rotation of the Earth.

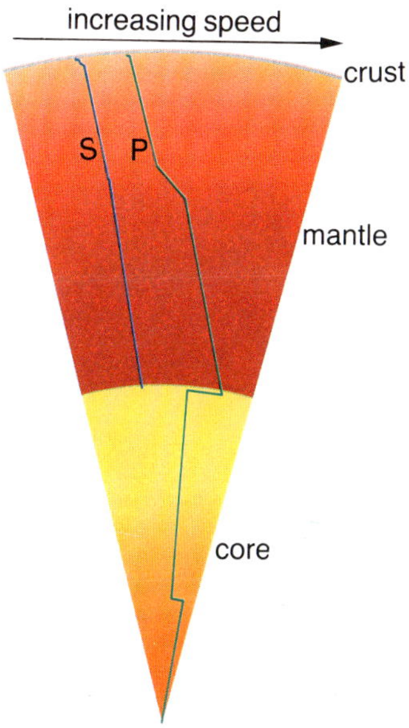

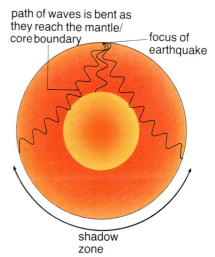

path of waves is bent as they reach the mantle/core boundary

focus of earthquake

shadow zone

Fig. 42.4   S waves travelling outwards from the focus of an earthquake never reach the far side of the Earth. This **shadow zone** on the far side of the Earth is caused by the inability of the waves to travel through the liquid core.

increasing speed

crust

S P

mantle

core

Fig. 42.3a   P waves suddenly slow down about 3000 km below the surface, S waves stop completely. This indicates a change from solid mantle to liquid core at this depth.

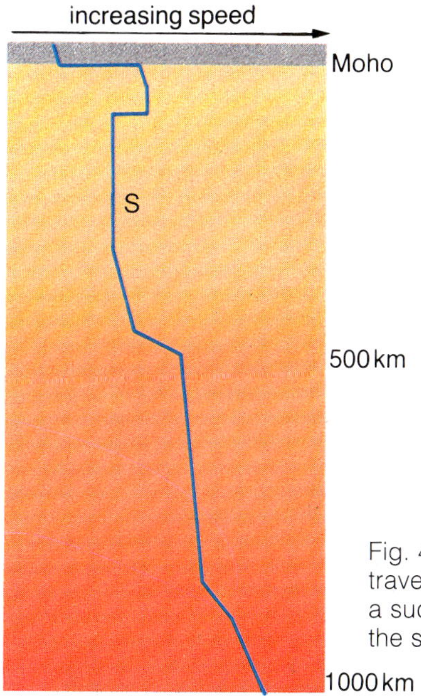

increasing speed

Moho

S

500 km

1000 km

Fig. 42.3b   A more detailed look at how S waves travel through the outer layers of the Earth shows a sudden increase in speed about 25–40 km below the surface. This is the Mohorovic discontinuity.

Questions

1 The deepest mine in the world is a gold mine in South Africa. It goes 3582 m below the surface.

a Into which layer of the Earth does this mine penetrate?

b What fraction of the Earth's diameter has been crossed by this mine?

2 Summarise the evidence which suggests that the Earth is made up of several concentric layers.

107

43 THE DRIFTING CONTINENTS

The continents are slowly drifting across the surface of the Earth.

The Earth's crust is made up of plates

The Earth's crust is not a continuous structure. It is made up of a series of **plates**. The plates behave rather like rafts, drifting across the surface of the mantle.

This may seem a rather strange idea! Our senses have no way of detecting this movement, because it is very, very slow. A map of the Earth drawn today would look very like one drawn a thousand years ago. The plates only move about 1cm a year.

No-one is quite sure what makes the plates move around. One suggestion is that convection currents are set up in the mantle. Some parts of the mantle are hotter than others. The hotter areas are less dense, so they move towards the surface. Here, they cool and become more dense. So they sink. These movements might cause friction between the mantle and the crust. They could make the plates of the crust move.

The study of the way these plates behave is called **plate tectonics**. Plate tectonics can help to explain many important features of the surface of the Earth, such as earthquakes, volcanoes and mountain formation.

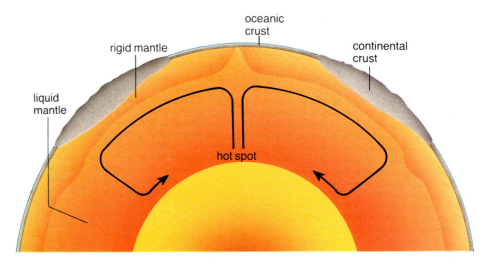

Fig. 43.1   Convection currents in the mantle produce forces which move the crust.

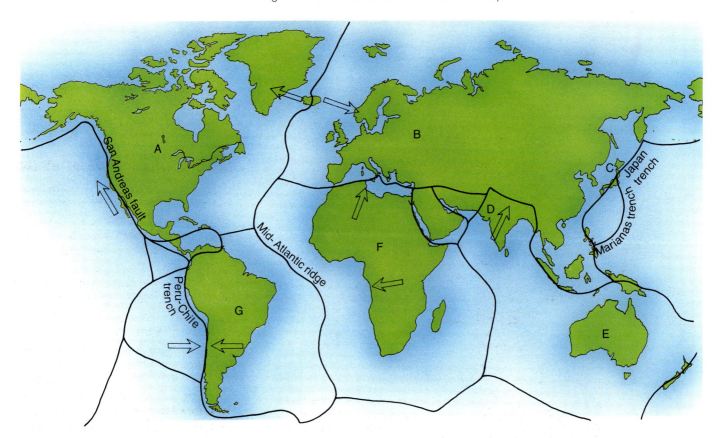

Fig. 43.2   World plate boundaries. The arrows show the direction of movement of the plates.

The continents once formed a single, huge land mass

If you look at a map of the Southern Hemisphere, it is not difficult to imagine the continents fitting together rather like jigsaw pieces. The most obvious fit is between South America and Africa. But Australia, Antarctica, New Zealand and India can also be fitted into the puzzle.

It is thought that, about 200 000 000 years ago, the continents were all joined together in one enormous land mass. Scientists have named this giant continent Pangea. Gradually, it began to break apart. First, the piece of land which was to become Europe broke away from the piece which was to become North America. Later, South America and Africa began to split apart. Between them, the Atlantic Ocean formed.

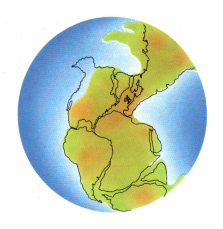

Fig. 43.3   Around 200 000 000 years ago, most of the land masses on Earth had moved together to form the giant super-continent, Pangea. Much of the interior of this huge land mass was hot and dry.

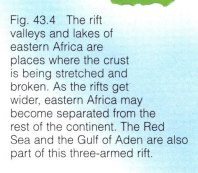

Fig. 43.4   The rift valleys and lakes of eastern Africa are places where the crust is being stretched and broken. As the rifts get wider, eastern Africa may become separated from the rest of the continent. The Red Sea and the Gulf of Aden are also part of this three-armed rift.

Red Sea

Gulf of Aden

Fig. 43.5   This fissure, called the Laki fissure, is in Iceland. It makes a boundary between two plates. It is a continuation of the boundary which runs down the centre of the Atlantic Ocean. The two plates are slowly moving apart.

The continents are still moving today

The continents are still moving. America is moving away from Europe and Africa. It is moving at about the same speed as the growth of your fingernails. So you are unlikely to notice much difference in your lifetime.

India is moving northwards, pushing into Asia. The forces generated by this movement are pushing up the Himalayan mountains.

Africa is swinging slowly clockwise. Eventually, it may swing right up to touch Spain, closing off the Straits of Gibraltar. This movement is also widening the Rift Valleys in the north and east of Africa. Geologists predict that these valleys will get wider and wider, eventually becoming new oceans.

There is a lot of evidence that the continents are moving across the surface of the Earth.

The continents seem to fit like jigsaw pieces

If you look at the shapes of America and Africa, you can see how they might fit together. Figure 44.1 shows a computer-generated best-fit picture of this. The fit was noticed many centuries ago, and several people suggested that perhaps America and Africa had once been joined together. But most geologists found it impossible to imagine that the continents could actually move across the surface of the Earth. Not until the early twentieth century was this idea taken seriously. And it was not until the 1960s that it was generally accepted.

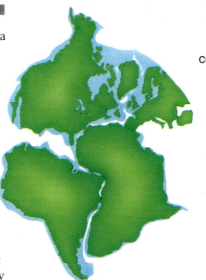

Fig. 44.1   The almost perfect jigsaw fit of the continents on either side of the Atlantic is strong evidence that they once formed a single land-mass.

continental shelf

land

Fig. 44.2   A fossil of the fern *Glossopteris*. Fossils of this fern have been found in South America, Africa, India, Antarctica and Australia, providing evidence that these continents were once joined together as one huge land-mass. This fossil was found in Australia, and is 260 000 000 years old.

Fossils suggest that the continents were once joined

In the early twentieth century, several scientists began to find other evidence that the continents may once have been joined together, and later have moved apart.

One of these pieces of evidence was the distribution of fossils in different continents. A group of fossil seed ferns, called *Glossopteris*, had been found in coal deposits in India, South Africa, Australia, South America and even Antarctica. The similarity between the fossils from these different places suggested that all these places might once have formed a single land mass, covered with similar vegetation. The seeds of *Glossopteris* are much too big to have been carried by the wind across the sea.

Animal fossils, too, suggested that the continents were once joined together. A fossil reptile, *Mesosaurus*, has been found in Brazil and in South Africa. It was a fresh-water animal, and could not possibly have crossed the Atlantic Ocean.

Fig. 44.3   Distribution of *Glossopteris*

areas where *Glossopteris* fossils have been found

Living organisms provide evidence for continental drift

Several groups of animals, still alive today, are also found in both America and Africa. One is a type of earthworm, found at the tip of South America and the tip of South Africa. It is hard to imagine an earthworm crossing the Atlantic Ocean. But if these two areas of land were once connected, then the distribution of this earthworm is much easier to explain.

Fig. 44.4 Marsupial mammals, such as these marsupial rats, are found only in Australia (except for opossums which are found in America). This can be explained if Australia was once part of a giant land mass but broke away at a time when marsupial mammals had evolved, but other mammals had not. In other parts of the world, more advanced mammals later evolved, and marsupials became extinct.

Rock sequences are similar in several continents

If rocks of similar ages are studied in Africa, America, Australia, India and Antarctica, strong similarities are found between them. For example, in South Africa, there are layers of rocks formed by material deposited by glaciers about 320 000 000 years ago. Rocks of similar age in Brazil, Antarctica and India were also formed from glacial deposits. On top of these rocks lie rocks formed from mud and plant remains, containing similar fossils. Above these rocks lie younger rocks, formed around 240 000 000 years ago. In all of these places, these younger rocks are sandstones, formed from drifting sand in a hot, dry environment. Above the sandstones are basalts, formed from solidified lava.

The similarities between the rock formations and sequences are so great that geologists are convinced that they formed when the continents were joined together as a single land mass.

Fig. 44.5 Rock sequences in four continents. The similarity between the kinds of rock formed at the same time suggests that they were also formed in the same place.

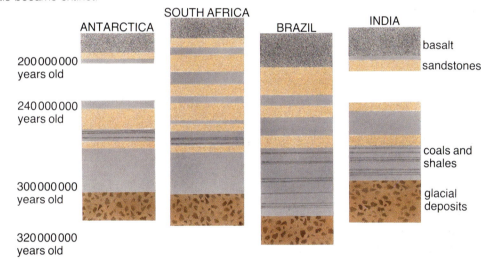

Particles in newly forming rocks line up with the Earth's magnetic field

When lava or magma solidify, their particles line themselves up according to the direction of the Earth's magnetic field. Once the rock is solid, the particles are 'frozen' in position. This alignment of particles can also happen when sedimentary rocks are forming.

Sandstones in Britain, formed about 200 000 000 years ago, have particles aligned in this way. But the way the particles are lined up is not what would be expected if the rocks had been formed where Britain is now. Their alignment can best be explained by suggesting that Britain has moved northwards thousands of kilometres in the last 200 000 000 years, and rotated 30° clockwise. Britain was once near the equator.

Care has to be taken when interpreting these results, because the Earth's North and South Poles reverse their magnetic polarity every now and then. The North Pole becomes a south pole, and the South Pole becomes a north pole. No-one really understands why this happens. The last change was about 1 000 000 years ago.

Plate tectonics can help to explain many geological events

Jigsaw fit, the distribution of fossils and living organisms, similar rock sequences and the alignment of magnetic particles in rocks all provide evidence that continents have drifted over the Earth's surface in the past. But perhaps the most convincing evidence for the theory of continental drift is the way in which it helps to explain so many features of the Earth's behaviour today.

In Topics 45 and 46, you will see how the movement of plates can help to explain where and why earthquakes and volcanoes happen, and how mountains and deep sea trenches can be formed. The correlation between what we see happening, and what we would expect to happen if continental plates really do move over the surface of the Earth, provides the firmest evidence of all that continental drift has happened, and is still happening now.

45 PLATE BOUNDARIES

Where two plates join, enormous forces may be generated. This can create earthquakes, volcanoes, mountains and trenches.

Plates are moving apart in the mid-Atlantic

Running the whole length of the Atlantic Ocean, roughly from north to south, is a junction between two plates. The plates are moving apart.

Hot material from the mantle rises up below this plate boundary. It spreads outwards, causing the plates to move apart. They buckle and bend, forming tall ridges. These ridges are like mountain ranges on the sea floor.

Some of the molten rock from the mantle escapes through cracks in the crust. The molten rock spreads outwards, solidifying on the sea bed. It forms new oceanic crust. Sometimes, the molten rock escapes with great force, forming undersea volcanoes. Some of these volcanoes are so tall that they reach above the surface of the sea. They form **volcanic islands**. This is how the Azores were formed.

The formation of new oceanic crust in the middle of the Atlantic, and the movement of the plates away from each other, is sometimes called **sea-floor spreading**. It is happening in the Pacific Ocean, too.

Fig. 45.1 Deep under the Pacific Ocean, at a boundary where two plates are moving apart, volcanic sulphide springs rise through the ocean floor.

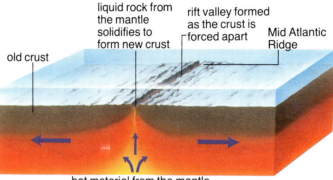

old crust · liquid rock from the mantle solidifies to form new crust · rift valley formed as the crust is forced apart · Mid Atlantic Ridge

hot material from the mantle rises upwards

Fig. 45.2 Sea-floor spreading at the Mid-Atlantic Ridge.

Trenches and mountains may be formed when two plates collide

As the oceanic plates move outwards, they collide with the continental plates. The oceanic crust is denser than the continental crust. So when they collide, the oceanic crust is forced below the continental crust.

Places where this happens are called **subduction zones**. A very active subduction zone runs along the west coast of South America.

As the oceanic crust is pushed below the continental crust, it becomes very hot. The rocks melt. As they melt, they expand. They push their way up to the surface. They form **volcanoes**.

The forces produced by these enormous masses of rock pushing under one another can generate **earthquakes**. Earthquakes happen when a force builds up to a point where one body of rock suddenly moves past another. Many of the world's earthquakes happen in subduction zones.

The continental crust above the subduction zone is pushed upwards. It folds and crumples, forming **mountain chains.** This is how the Andes were formed.

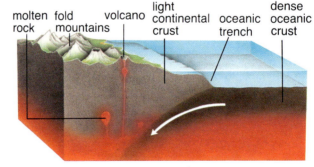

molten rock · fold mountains · volcano · light continental crust · oceanic trench · dense oceanic crust

Fig. 45.3 The huge forces generated when two plates collide build mountains and form explosive volcanoes.

Fig. 45.4 The Andes mountains

Collisions between continental plates can form fold mountains

If two plates of continental crust collide, a subduction zone is not formed. This is because the two plates are of the same density. So one does not sink deeply under the other. But if they keep moving towards each other, one may push under the other.

An example of two continental plates colliding is the movement of India into Asia. India has moved northwards, ramming into Asia and pushing beneath its surface. The force of this collision has pushed up the Himalayas.

Another mountain range thought to have formed by such a collision is the Alps. Africa and Europe, revolving slowly, pushed into each other. The enormous forces made the crust fold and buckle. Mountains formed in this way are called **fold mountains**. The Andes, the Alps and the Himalayas are all examples of fold mountains. They tend to form in long chains, all the way along the boundaries between moving plates.

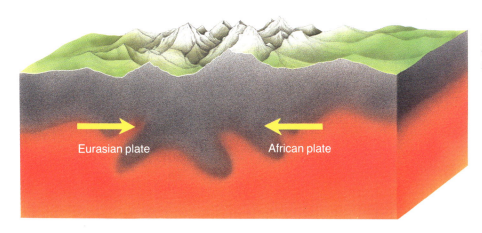

Eurasian plate　　African plate

Fig. 45.5   The Alps were formed when two plates collided around 50 000 000 years ago.

Plates moving past one another can cause earthquakes

Some plates are moving sideways past one another. One place where this is happening is in California. Here, the Pacific Plate is sliding past the North American Plate. The boundary between them forms the San Andreas Fault.

The plates do not move smoothly and steadily. Friction stops them moving. Then the forces build up to such a strength that the plates suddenly jerk past each other. This causes an earthquake.

There have been several large earthquakes in California caused by movement along this plate boundary. A major earthquake occurred in 1989. There are certain to be more. San Francisco lies close to the plate boundary. Many of the buildings in San Francisco are built to be able to withstand earthquake damage. But a major earthquake would almost certainly cause widespread destruction and loss of life.

Fig. 45.6   The 1989 San Francisco earthquake tore roads apart as the ground moved beneath them.

Questions

1 a What are fold mountains?
  b Give two examples of fold mountains.
  c Describe how fold mountains may be formed.
2 a What is a subduction zone?
  b Explain why many of the world's earthquakes and volcanoes occur near subduction zones.
3 An American ship, the *Glomar Challenger*, drilled a series of holes in the floor of the Atlantic Ocean. Rocks from the holes were analysed, and their ages measured. The results for one set of holes are shown in the table on the right.

Distance from Mid-Atlantic Ridge (km)	Age of rock sample (millions of years)
200	10
420	24
520	27
750	32
860	40
1000	48
1300	66
1680	78

  a Plot a graph to show these results. Distance should go on the horizontal axis.
  b What is the Mid-Atlantic Ridge?
  c Suggest one way in which the geologists might have dated their rock samples.
  d How do these results support the theory of sea-floor spreading?
  e According to these results, about how fast is the sea-floor spreading outwards?

46 VOLCANOES

A volcano is a gap through which hot, molten rocks can escape to the surface. Different kinds of molten rock form different kinds of volcanoes.

Volcanoes may form at plate boundaries

Beneath the Earth's crust, there are areas where the rock is so hot that it is liquid. This molten rock tends to be found near plate boundaries. The molten rock is called **magma.**

A volcano is a place where magma escapes to the surface of the Earth. When the magma reaches the surface, it is called **lava**.

Not all volcanoes form at plate boundaries. Some are formed at other 'hot spots' on the Earth. The Hawaiian islands, for example, are volcanic, but they are not at a plate boundary.

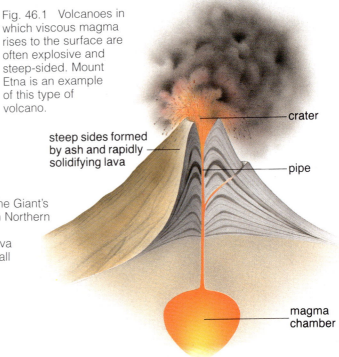

Fig. 46.1 Volcanoes in which viscous magma rises to the surface are often explosive and steep-sided. Mount Etna is an example of this type of volcano.

crater

steep sides formed by ash and rapidly solidifying lava

pipe

magma chamber

Fig. 46.2 The Giant's Causeway in Northern Ireland was formed as lava cooled into tall columns of basalt.

Colliding plates tend to form explosive volcanoes

Volcanoes are formed when two plates collide. You will remember that when oceanic crust collides with continental crust, the oceanic crust is forced down, underneath the continental crust. The oceanic crust, and all the mud and other sediments on top of it, is carried deep below the continental crust. The huge pressures and temperatures melt the oceanic crust and sediments. They form magma.

This magma is very thick, and does not flow easily. It is **viscous**. It rises slowly towards the surface. As it rises, it cools and may solidify. It may block the pipe of the volcano

(see Figure 46.1). Enormous pressures build up. Gases may become trapped in the magma. If the gases suddenly expand, then the volcano erupts violently.

The material which erupts from the volcano is often a mixture of **ash** and **hot gases**. Usually, only a little **lava** flows out. The clouds of hot ash and lava flow downwards, forming a steep-sided cone. The lava solidifies quickly, so it does not have time to spread out far before it becomes solid rock.

Fissure eruptions form where plates are moving apart

Where plates are moving apart, a different sort of volcano may form. In these areas, the magma beneath the solid crust is more runny than the viscous lava which forms explosive volcanoes. This runny lava seeps more gently out of fissures. The lava takes a while to solidify, so it has time to run over the surface for quite large distances. It forms **lava plateaus**. If the volcano is active for a long time, then layer upon layer of lava may build up.

Some lava plateaus are enormous. In Antrim, in Northern Ireland, there is a lava plateau covering 150000 km². It was formed about 50000000 years ago. It was much

bigger than this when it was first formed, but a lot of it has since disappeared beneath the Atlantic Ocean as the plates have continued to move.

Fig. 46.3 Volcanoes in which 'runny' (less viscous) lava rises to the surface are not explosive. Mauna Loa in Hawaii is an example. The lava often leaks out through long cracks or fissures.

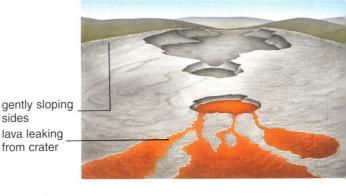

gently sloping sides

lava leaking from crater

Volcanoes can cause tremendous damage

A volcanic eruption, even a gentle one, can be a very dangerous event. The Mount Saint Helens eruption in 1980 is a good example of the different ways in which volcanoes can affect life and landforms around them.

Mount Saint Helens lies near the boundary between the Pacific Plate and the North American Plate. In May 1980, there was a small earthquake in this region. On May 18, Mount Saint Helens suddenly erupted.

Very little lava flowed out. Most of the material was dust and ash. There was about 400 000 000 tonnes of it. Some of it rolled down the mountain-side, killing 61 people and countless animals. The force of the explosion felled thousands of square kilometres of forest. The explosion was the equivalent of that produced by a 10 000 000 tonne bomb.

Some of the dust and ash was blown as high as 18 km. A huge dust cloud formed, which spread out over North America and the Atlantic Ocean. The cloud remained in the air for many months.

The dust and ash which settled around the mountain covered 260 000 km^2, to a depth of about 2 mm. It covered plants and killed them.

The heat from the ash melted snow and ice on the top of the volcano. The melted snow and ice flowed down the mountain, mixing with the ash to form mud-flows. The mud-flows blocked rivers and roads, flattened forests and destroyed houses.

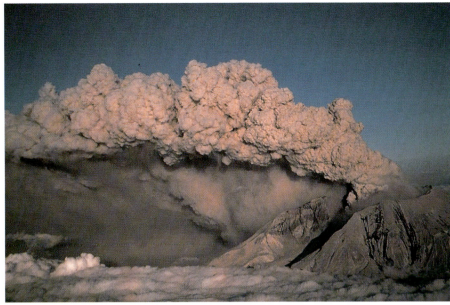

Fig. 46.4   A huge column of gas, ash, and rock particles erupted from Mount Saint Helens in July 1980. Most of the living things on the mountain were killed in the eruption, but some, such as gophers, survived underground.

Fig. 46.5   Although 61 people were killed in the Mount Saint Helens eruption, this is a very small number compared with the 22 000 killed when the Nevada del Ruiz volcano in the Andes erupted in 1985. Most of these people were killed by mud flows. The mud flows were caused when the eruption melted ice and snow on the top of the volcano. The melted ice and snow mixed with the ash and dust and flowed down the mountain-side at speeds of up to 100 km/h. There had been warnings that the volcano might erupt but people were reluctant to leave their homes and many had nowhere else to go.

Fig. 46.6   The rich soil on the slopes of Mount Bromo in Java allows the cultivation of crops like cabbages and onions.

Soil around volcanoes is very fertile

Despite the dangers of volcanoes, many people choose to live near them. One reason is that the ash from volcanoes contains many nutrients, such as potassium, which are needed by plants. So crops grow well in soil formed from volcanic ash.

Many volcanoes are **dormant**. This means that they have not erupted for a very long time. But even dormant volcanoes can eventually erupt again. Often, people have some warning that this is going to happen, and can leave their homes in time.

47 IGNEOUS ROCKS

Igneous rocks form when magma or lava cools and solidifies. They sometimes contain valuable ore deposits.

Magma and lava form crystals as they cool

You will remember that liquid rock is called magma when it is beneath the Earth's surface, and lava when it emerges from the ground. When magma or lava cools, it solidifies. It forms solid rock. Rocks formed in this way are called **igneous rocks**.

Magma and lava are made up of a variety of **silicates**, and dissolved gases, especially **steam**. Silicates are compounds made from silicon and oxygen, together with smaller amounts of other elements such as iron, magnesium and potassium.

As the liquid rock cools, these minerals form crystals. Igneous rocks are crystalline rocks. The size of the crystals depends on how the rock crystallises. **Granite**, for example, is formed when magma crystallises underground. The magma cools slowly, so large crystals form. **Basalt** is formed when lava solidifies above ground. It cools quickly, so the crystals formed are much smaller than those in granite.

Fig. 47.1 Basalt is a fine-grained rock, made of tiny crystals of several kinds of silicate. This sample is rather unusual, as it also contains large crystals of the minerals olivine and pyroxene (magnesium and iron silicates).

Fig. 47.2 Granite is also mostly made of silicates, but the crystals are larger than those in basalt. The large pink crystals in this sample are crystals of a mineral called orthoclase, which is a potassium aluminium silicate.

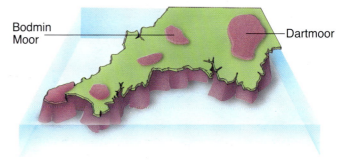

Bodmin Moor — Dartmoor

Fig. 47.3 An enormous mass of granite (shown in purple) lies under much of Cornwall and Devon. It formed about 250 000 000 years ago, as a giant blister of molten rock solidified underground. In places, the overlying rock layers have worn away, so that the granite is now at the surface.

Fig. 47.4 Granite is often used for building, or for making steps or pavements. It is a very hard rock, but it occurs naturally in blocks, making it easier to work with. It is very resistant to temperature changes, and to rain, and is hard wearing. Its variety of different crystals gives it an attractive mottled appearance.

INVESTIGATION 47.1

Crystallisation

As magma cools, crystals form. You are going to melt a powder, and then watch crystals form as it solidifies.

1 Set up a microscope, using the lowest power objective lens.

2 Put a small amount of salol powder onto a clean microscope slide. Holding the slide in your fingers, warm it gently in a Bunsen flame. Don't get it too hot. The salol will melt.

3 Put the warm, *not hot*, slide onto the stage of the microscope. Watch what happens as the salol solidifies. If you like, you can melt the salol again.

4 Record your results carefully. A drawing is probably the best way of doing this.

Ores may be concentrated as igneous rocks form

As magma solidifies and crystallises, some minerals may become separated from others. Sometimes, for example, certain minerals may dissolve in hot water within the cooling rock. As the hot water collects within the rock, or seeps into the other rocks around it, these dissolved minerals are carried along with it. Eventually, the water escapes, leaving the minerals behind as crystals. These minerals may contain metal compounds. They are called **ores**.

The shapes of rocks give clues about their formation

You have seen how the crystals in igneous rocks tell you something about how the rocks formed. Other clues can be found in the shapes of the rocks.

For example, magma cooling underground may seep in between layers of older rocks, forming a **sill**. Or it may seep upwards through a weak place in the rocks around it, to form a **dyke**.

Lava cooling on the surface also makes particular patterns. Wrinkles may form on the surface of thick, viscous

Fig. 47.5 The granite of Cornwall and Devon formed deep underground, but now some is exposed at the surface, because the overlying rocks have worn away. This is Vixen Tor on Dartmoor in Devon. The cracks in the granite have appeared because the rock was originally formed under tremendous pressures, which were released as the overlying rocks disappeared.

Fig. 47.6 Cornwall has ore deposits of tin, tungsten and lead. These ores were formed as the solidifying magma cooled within the surrounding rocks. Mining used to be an important industry in Cornwall, but it has now become uneconomical.

granite
ores

lava as it cools. Lava which is not so thick may form wide, flat areas of smooth rock. If lava erupts under the sea, and cools under water, it may form a rock with a hummocky surface, called **pillow basalt**.

Fig. 47.7 A vein of copper ore in granite in Cornwall

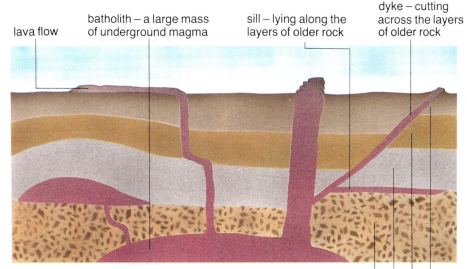

lava flow

batholith – a large mass of underground magma

sill – lying along the layers of older rock

dyke – cutting across the layers of older rock

layers of older sedimentary rocks

Fig. 47.8 Magma may seep between cracks in older rocks before it cools and solidifies. If these older rocks are later eroded away, the solidified magma becomes exposed.

Fig. 47.9 This rock was formed from flowing lava, which wrinkled as it cooled and hardened.

Sedimentary rocks form when solid particles collect together and then solidify. These rocks often contain fossils.

Rocks may form from particles of older rocks

Rock which is exposed on the surface of the Earth is constantly being worn away by water and wind. This is called **weathering**. You will find more about weathering in Topic 52.

Weathering breaks rock down into particles. These particles may then be carried away. They may simply fall down a hillside under the influence of gravity. They may be blown away by the wind. Or they may be carried away in streams and rivers, or in a slowly moving glacier.

The rock particles collect together in another place. For example, a river may carry mud into a lake. In the lake, the mud gradually sinks to the bottom. It forms a **sediment**. The process continues, perhaps for hundreds or thousands of years. The mud is pressed down under the weight of more mud sedimenting on top of it. The lower layers are pressed together, or **compacted**. Eventually, they are pressed together so solidly that they form solid rock.

Different sized particles form different kinds of rock. Very small particles form the fine-grained rock **shale**. Larger, sand sized grains, form **sandstone**. Gravel sized particles form **conglomerates**.

Fig. 48.1   Sediment builds up in layers at the bottom of a lake. The older layers are below the newer ones. After thousands of years, the sediments turn to rock.

Fig. 48.2a   **Shale** is a sedimentary rock formed from tiny mud particles. The particles collected in layers, and in this sample the layered structure is allowing the rock to split.

Fig. 48.2b   **Sandstone** is formed from sand grains. This sample of sandstone was formed from wind-blown sand, in a desert.

Fig. 48.2c   This **conglomerate** is made of large, rounded pebbles of flint embedded in a matrix (background) of smaller particles. Rocks like this are sometimes called 'pudding stones'.

Sedimentary rocks can form on land or under water

There are many different ways in which sedimentary rocks can form. As well as sedimentation in a lake, **rivers** may deposit mud, silt and gravel on a flood plain. Over hundreds of thousands of years, the sediments turn into rock.

Glaciers also deposit rock particles and soil. In Britain, **tills** are made up of sedimentary rock formed by material left behind by glaciers at the end of the last ice age.

Silt and mud can also be deposited at the bottom of the **sea**. Here, too, huge thicknesses of shells from tiny marine animals may build up. These shells, made from calcium carbonate, slowly become squashed so tightly together that they form rock. The rock is **limestone**.

Limestone can also form when there is so much calcium carbonate dissolved in sea water that it begins to precipitate out of solution. This is most likely to happen in warm water.

Limestone and sandstone are used as building materials

Limestone makes an excellent building stone. It is a **freestone**, which means that it can be split in any direction. This makes it easy to shape into blocks for building walls. But the purest type of limestone, chalk, is too soft to use for building, and dissolves too easily in rain water.

Sandstone, too, is often used for building. It, too, can be shaped into blocks.

Limestone has another important use in building. Roasted limestone is used for making **cement**. Usually, chalk is used, because it is the purest type of limestone, and is soft and easy to work. Clay and small amounts of gypsum are mixed with the roasted limestone, to help it to set.

Sedimentary rocks may contain fossils

As sediments collect on land or under water the bodies of animals or plants may fall into them. If more and more sediments are deposited, then the bodies become buried. They may form **fossils**. The fossils become trapped inside the layers of sediment, which eventually form sedimentary rock.

Only sedimentary rocks contain fossils. Igneous rocks never contain them.

Fig. 48.3   A sample of limestone made from the shells of animals which lived in the sea 400 000 000 years ago, when deep, warm water covered most of Britain.

Fig. 48.4   Limestone is widely used as a building material. It was used to build Tower Bridge in London. Its light colour is very attractive. It is a freestone, which means that it can be cut in any direction. It can, however, be very easily damaged by acid rain, because it contains calcium carbonate which reacts with acid and dissolves.

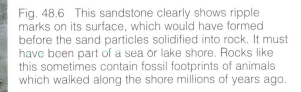

Fig. 48.5   The different layers of sediment which have built up on one another can clearly be seen in this rock on Ellesmere Island in the Canadian Arctic.

Fig. 48.6   This sandstone clearly shows ripple marks on its surface, which would have formed before the sand particles solidified into rock. It must have been part of a sea or lake shore. Rocks like this sometimes contain fossil footprints of animals which walked along the shore millions of years ago.

Fig. 48.7   A dinosaur fossil from New Mexico. The body of this bird-like dinosaur became trapped in soft sediments in a lake. The soft tissues decomposed, but the bones fossilised and were preserved in the sediments as they turned to rock. The rock has been chipped away from the fossilised bones to show them more clearly.

49 METAMORPHIC ROCKS

Very high temperatures and pressures can change the structure and composition of rocks.

Sedimentary and igneous rocks can be altered

The mineral structure and form of rocks can change or **metamorphose**. This can happen if rock is heated by contact with magma. It can also happen if rock is subjected to great pressures underground.

Any kind of rock can be metamorphosed. Clays and shales, which are sedimentary rocks, can be pressed and squeezed so greatly that the minerals from which they are made change their structure and form. Shale becomes **slate**.

Another kind of sedimentary rock, limestone, may be changed to **marble**. Very high temperatures break up the shells and other material in the limestone. They reform as small, even-sized crystals, which make the rock much smoother and harder than before.

Metamorphism often happens to rock which is buried deep underground. Here, the high temperatures and pressures may operate over very large areas – perhaps hundreds of kilometres. So large quantities of rock are metamorphosed. This is called **regional metamorphism**.

Metamorphism can also be caused when magma comes into contact with other rocks. Rock near to the magma is heated and metamorphosed. These changes happen around the edges of the hot magma. This is called **contact metamorphism**.

CLAY

Clay is made up of tiny, plate-like particles, lying in all directions.

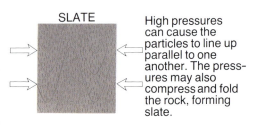

SLATE

High pressures can cause the particles to line up parallel to one another. The pressures may also compress and fold the rock, forming slate.

Fig. 49.1 How slate is formed from clay

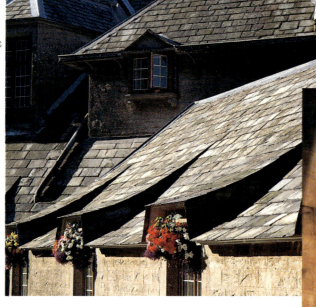

Fig. 49.2 Slate is a very attractive traditional roofing material. However, it has now become expensive, and is also very heavy, requiring a sturdy timber roof to support it. Cheaper and lighter artificial substitutes are now often used instead.

Slate and marble are used for building and decoration

Slate is used for roofing. As the clay and shale from which it forms are metamorphosed, flaky minerals in it become lined up in one direction. Slate can be split in this direction, into thin, flat sheets. But water cannot seep through it from one side to the other, because it would have to go through these layers. So slate is ideal for making thin, flat, hard, waterproof roofing slates.

Marble is hard, and has an even texture. It can be polished, often showing attractive patterning. So marble is used for decorative parts of buildings.

Fig. 49.3 Marble has been used decoratively in buildings for centuries. This is the Palace Theatre in London.

Radioactivity and fossils can be used to find the age of rocks

Geologists often want to be able to compare the ages of rocks in different areas, sometimes in different countries or even continents. The earliest way of doing this was to look at similarities in rock sequences. Figure 44.5 shows one example of this. If two rock sequences are very similar, then they may have been formed at about the same time.

Another, much better, method is to look at **fossils** in the rocks. Sedimentary rocks which formed at similar times, and in similar environments, will contain similar fossils. So if you find two layers of rock with identical fossils, you can be sure that the rock in these layers are the same age. But it does not always work the other way round! If the two layers of rock contain *different* fossils, they may still be the same age. They may have been formed in different environments, where different animals and plants lived.

Neither fossils nor rock sequences tell you the *actual* age of a rock in years. They only tell you the *relative* ages of the rocks. The actual age of a rock can only be given by a dating method which uses **radioactive isotopes**. Thorium 232, for example, is used to date very old rocks which are thought to be over 50 000 000 years old. Thorium 232 decays to lead 208 and has a half life of 13 900 000 000 years. By comparing the amount of thorium 232 in a rock with the amount of lead 208, you can work out the age of the rock.

Fig. 50.3 The main rock layers in the Grand Canyon. See Table 50.1 for the ages of the periods.

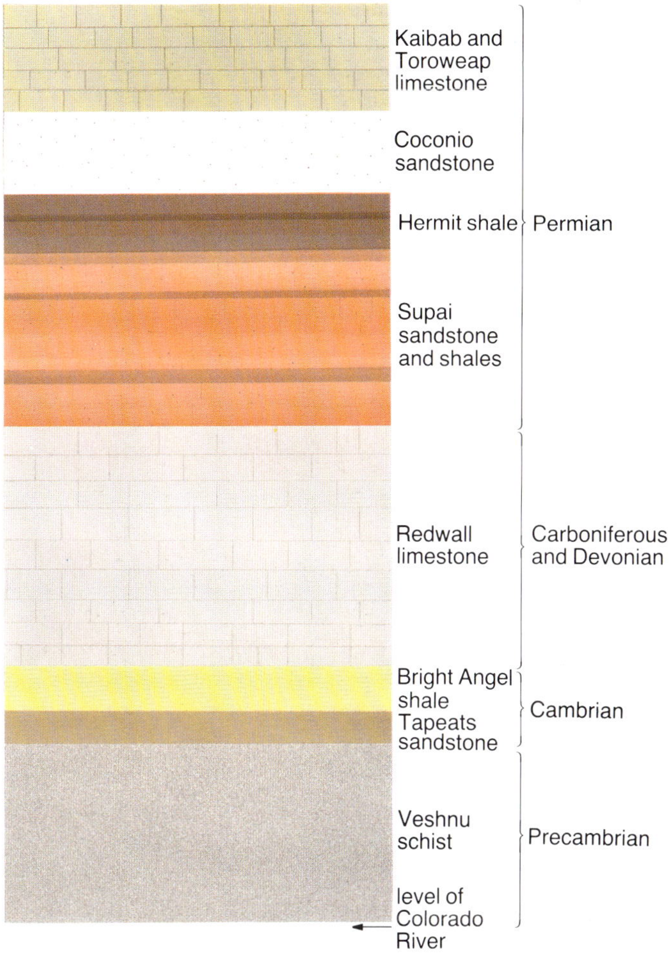

period in which rock was formed

Kaibab and Toroweap limestone

Coconio sandstone

Hermit shale — Permian

Supai sandstone and shales

Redwall limestone — Carboniferous and Devonian

Bright Angel shale
Tapeats sandstone — Cambrian

Veshnu schist — Precambrian

level of Colorado River

Fig. 50.4 The Grand Canyon in Colorado, USA. The Colorado River has, over thousands of years, cut down through layer after layer of rock forming a canyon almost 350 km long, between 6 and 20 km wide and 1615 m deep. The layers are a slice through the Earth's history. The oldest exposed rocks at the bottom of the canyon are Precambrian. The youngest rocks, at the top, are Permian.

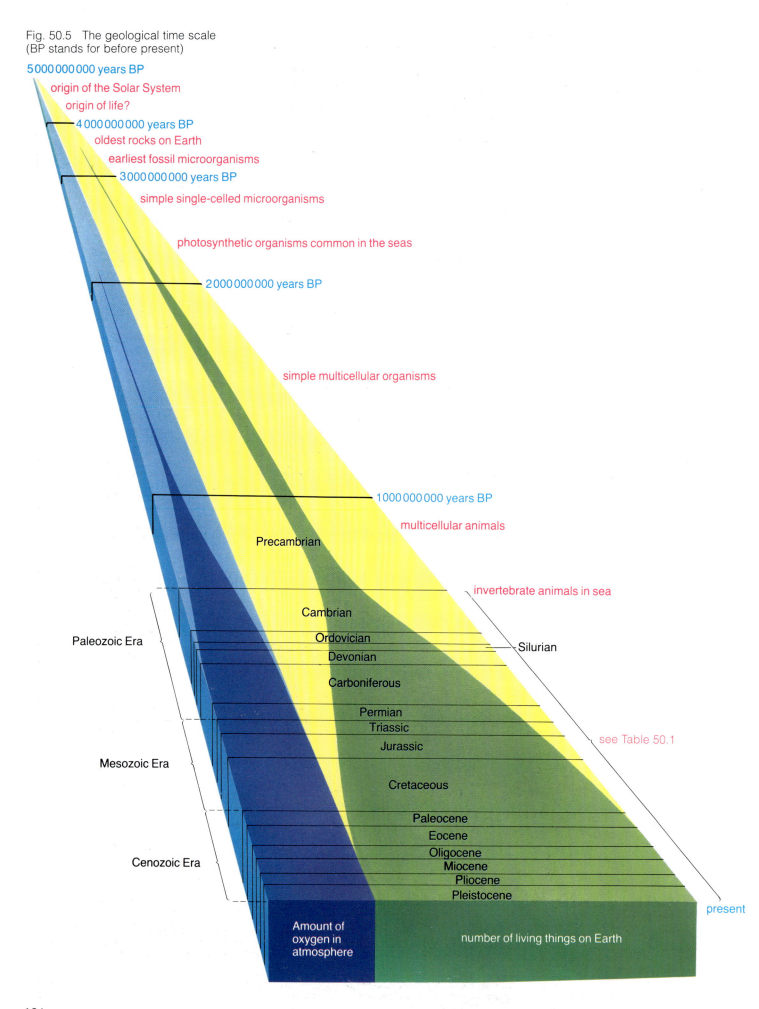

Fig. 50.5 The geological time scale
(BP stands for before present)

5 000 000 000 years BP

origin of the Solar System

origin of life?

4 000 000 000 years BP

oldest rocks on Earth

earliest fossil microorganisms

3 000 000 000 years BP

simple single-celled microorganisms

photosynthetic organisms common in the seas

2 000 000 000 years BP

simple multicellular organisms

1 000 000 000 years BP

multicellular animals

Precambrian

invertebrate animals in sea

Cambrian

Paleozoic Era

Ordovician

Silurian

Devonian

Carboniferous

Permian

Triassic

Jurassic

see Table 50.1

Mesozoic Era

Cretaceous

Paleocene

Eocene

Cenozoic Era

Oligocene

Miocene

Pliocene

Pleistocene

present

Amount of oxygen in atmosphere

number of living things on Earth

Table 50.1   The geological time scale

Millions of years ago	Name of period or epoch	Conditions in Britain	Examples of living things	World events
0.5 —				
	Pleistocene	Fluctuating climate, alternating between warm and very cold.	Neanderthal man and woolly mammoths.	Ice age until 10 000 years ago.
1.8 —				
	Pliocene	Initially warm, but then cooling.	Humans first appear.	Sea levels rise, and climate gets warmer. Ice age begins 3 000 000 years ago.
5 —				
	Miocene	Oaks, beeches and redwoods grow. There are expanses of grassland.	Cooler drier climate allows grasses to cover large areas.	Climate continues to get colder and drier. Ice sheets expand in Antarctica, so sea levels drop. The Mediterranean dries up.
24 —				
	Oligocene	Britain gradually drifts northwards.	Sabre-toothed cats and monkeys appear.	The Alps are formed. Climate becomes cooler and drier.
37 —				
	Eocene	Warm, almost tropical climate. Crocodiles live in the sea.	Increase in variety of mammals, including bats, whales and elephants.	Australia breaks away from Antarctica and moves northwards.
58 —				
	Paleocene	Intense volcanic activity in Northern Ireland and Scotland.	First primates and horses appear.	
65 —				
	Cretaceous	Covered by warm seas. Limestones of the Downs deposited.	Flowering plants appear. Many insects. Dinosaurs become extinct.	High sea levels. Continents continue to move further apart.
140 —				
	Jurassic	Changing pattern of land, seas, swamps and lakes. Britain lies 40° north of the equator.	Dinosaurs abound.	Atlantic Ocean opens between Europe and North America.
200 —				
	Triassic	Seas spread over the land.	Age of dinosaurs begins. First birds and mammals appear.	Pangea begins to break up.
240 —				
	Permian	Covered by desert.	Many invertebrates and vertebrates become extinct at the end of the Permian.	Continents continue to move together, forming the super-continent Pangea.
280 —				
	Carboniferous	Southern Britain lies on equator. Warm seas gradually withdraw, and coal swamps flourish on land.	Reptiles and amphibians on land. Primitive trees form forests. Winged insects.	Widespread coal swamp. Continents moving together.
350 —				
	Devonian	Britain lies near equator. Sea covers the south west. On land, a hot wet climate.	First insects appear. Fish are still the only vertebrates. Primitive trees form forests.	Granites form in south west Britain, Scotland and Wales.
400 —				
	Silurian	Deep water covers much of Britain. Corals grow in shallow areas.	Green plants and arthropods invade the land. Ammonites appear. First jawed fish appear.	High sea levels and warm climate.
440 —				
	Ordovician	Britain lies just south of the equator, mostly covered by sea. Much volcanic activity.	Many different invertebrates live in the sea.	Shallow seas cover most of the continents.
500 —				
	Cambrian	Britain covered by sea.	Trilobites and other shelled invertebrates roam the sea. Jawless fish appear.	Most continents lie south of the equator. Sea levels are high.
570 —				
	Precambrian	The Precambrian era includes almost 90% of geological time. The Earth was first formed about 4 500 000 000 years ago. The earliest traces of life, very simple single-celled bacteria, are about 3 500 000 000 years old. At first, the atmosphere contained very little oxygen. Fossils of simple soft-bodied invertebrates first appeared 1 000 000 000 years ago.		

51 FOLDING AND FAULTING

Layers of rock may be twisted, split and folded by Earth movements.

Mountain building involves tremendous forces

In Topic 45, you saw how fold mountains may be formed. When two plates collide, they push the crust upwards. Most of the Earth's mountain ranges have been formed in this way.

Sedimentary rocks are laid down in layers, with the youngest rocks on top. Igneous rocks may also form part of these layers. The huge forces involved in mountain building can put these layers of rock under tremendous pressure. The layers may be folded or split.

The south-east of Britain does not contain any mountains but it does have the Downs. These hills were formed when the African plate collided with the plate on which Britain was situated, about 25 000 000 years ago. Before the collision,

there was a flat plain, covered by a 300m layer of chalk. Under the chalk were layers of clay and sandstone. The forces of the collision **folded** the rocks.

A long dome of rock was pushed up. This dome of rock is called an **anticline**. It forms the crests of the Downs. Since the anticline was pushed up, its top has been gradually eroded away. Now the Downs are gentle, rolling hills.

A long hollow was also formed. This hollow is called a **syncline**. The syncline was flooded by a sea 60 000 000 years ago. Mud was deposited under the sea, forming clays. Now the sea has gone, and the filled-up syncline forms the London Basin.

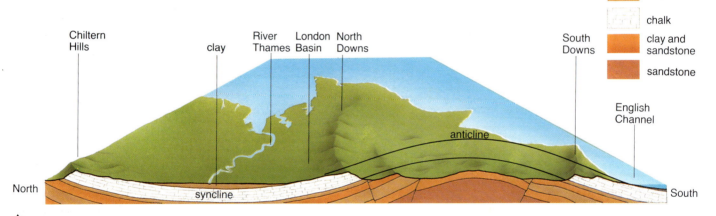

clay

chalk

clay and sandstone

sandstone

Fig. 51.1 The North and South Downs, in southern Britain, were formed by folding followed by erosion. The top of the fold, called an anticline, has eroded away, leaving steep chalk slopes or scarps. The bottom of the fold, called a syncline, has filled with clays deposited by the sea, which flooded the low-lying land.

Fig. 51.2 The South Downs from Bury Hill. Notice the chalky soil, the rolling hills and the exposed chalk face in the middle distance.

Fig. 51.3 The layers in these rocks would originally have been horizontal. Plate movements have lifted and folded them. This is Stair Hole near Lulworth in Dorset.

Rock layers may be split apart

Another way in which rock layers may react to huge forces is by splitting or **faulting**. Faulting often occurs at plate boundaries.

In the East African Rift Valley, plates are being pulled apart. As the crust splits apart, cracks develop. The land in between the cracks drops down. In East Africa, this has resulted in the formation of a huge valley, with mountains on either side.

Faults can also develop where other types of crust movement are taking place. Figure 51.4 shows some of them.

Fig. 51.4   Types of fault. The arrows show the forces which cause the rock to move.

Sedimentary rock forms in layers.

A **tear** fault is formed as two plates move past each other.

A **normal** fault is formed as the crust is pulled apart. The surface is lengthened.

A **reverse** fault is formed as the crust is compressed. The surface is shortened.

Fig. 51.5   A cross-section through a rift valley. As the crust is pulled apart, a series of faults forms. The central block of land drops down, leaving mountains on either side of a valley. Magma can be forced up to the surface along the fault lines, forming volcanoes.

▼

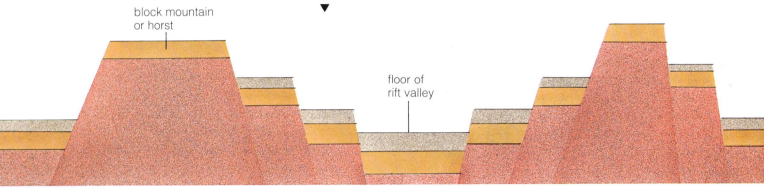

block mountain or horst

floor of rift valley

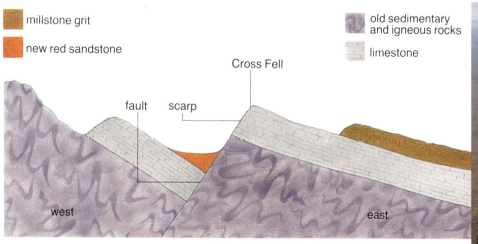

millstone grit

new red sandstone

old sedimentary and igneous rocks

limestone

Cross Fell

fault   scarp

west

east

Fig. 51.6   Faulting in the Pennines at Cross Fell. The fault has produced a long, steep slope, called a scarp. What type of fault is it? Which are the youngest rocks in the diagram? How can you tell?

Fig. 51.7   The San Andreas Fault runs along almost the whole length of California. It marks the boundary between the Pacific Plate and the North American Plate, which are sliding past each other. The ridges on either side of the fault have been formed by the huge pressures generated by the fault movements.

Questions

1 a Under what conditions might the layer of chalk, from which the Downs were formed, have built up?

b Find the Chilterns, South Downs and North Downs in your atlas. Draw a map to show their position.

c Figure 51.1 shows the relative positions of some of the layers of rocks in southern Britain as they are today. Which are the youngest rocks shown in this diagram? Which are the oldest?

d What evidence might geologists have used to decide that the chalk of the Chilterns, the North Downs and the South Downs was all laid down in the same place at the same time?

Weathering is the breakdown of exposed rock

Many rocks, such as granite, are formed deep underground. But as the ground above them is gradually worn away, they may become exposed at the Earth's surface. Here they are subjected to the action of air, water and changing temperatures. They are slowly broken down. This is called **weathering**.

Physical weathering is caused by temperature changes

Rain water soaks into cracks in rocks. If the temperature drops below 0 °C, the water freezes. As it freezes, it expands, forcing the rock apart on either side of the cracks. When the ice melts, the rock may split along the cracks. This is especially likely to happen in rocks where there are already lines of weakness, such as in sedimentary rocks like shales, limestones and sandstones.

This kind of weathering is important in temperate countries like Britain, where alternate freezing and thawing happens a lot. In mountainous regions, the pieces of broken off rock fall down the mountain, forming **scree**.

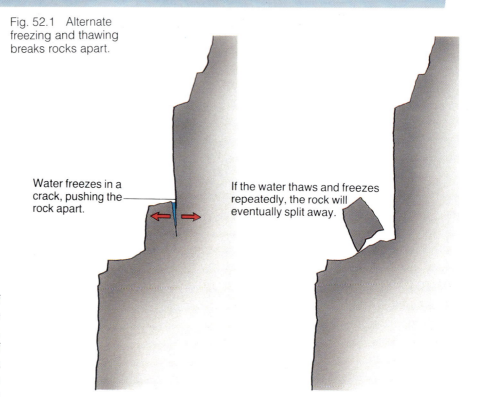

Fig. 52.1 Alternate freezing and thawing breaks rocks apart.

Water freezes in a crack, pushing the rock apart.

If the water thaws and freezes repeatedly, the rock will eventually split away.

Fig. 52.2 Screes at Wast Water in the Lake District. Wast Water is Britain's deepest lake.

Acids can dissolve rock

Rain water is a weak acid. This is because it contains small amounts of dissolved carbon dioxide. Over long periods of time, the acid can react with limestone and dissolve it.

carbon dioxide + water + limestone (calcium carbonate) $\longrightarrow$ calcium hydrogencarbonate

$$CO_2(aq) + H_2O(l) + CaCO_3(s) \longrightarrow Ca(HCO_3)_2(aq)$$

Calcium hydrogencarbonate is soluble in water. It is washed away as water flows over and through the limestone. The effects can be quite dramatic, producing unique landscapes and scenery in limestone regions. Underground caves are often formed in the limestone.

Acid rain, formed when oxides of sulphur and nitrogen dissolve in water, has an even lower pH than 'normal' rain. It can dissolve limestone much faster.

INVESTIGATION 52.1

The effect of acid on rocks

1 Take small samples of several different rocks. Place them in a glass petri dish, or on a tile. Make sure you know which one is which.

2 Using a dropping pipette, put a few drops of dilute hydrochloric acid onto each sample. Record what happens immediately after adding the acid, and what happens when you leave the samples for a day or so.

Questions

1 Which rocks were affected by the acid? Which were not affected?

2 Suggest why some rocks were affected more than others.

3 Hydrochloric acid does not usually fall onto rocks. But rocks are affected by acid. Where does the acid come from?

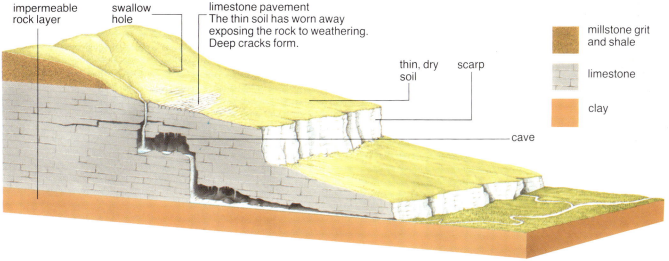

Fig. 52.3   Limestone scenery. Limestone is permeable to water, so water moves down through it, leaving the ground surface dry. The water dissolves the rock as it flows through, forming underground passages and caves. This sort of landscape is known as **karst**.

impermeable rock layer

swallow hole

limestone pavement
The thin soil has worn away exposing the rock to weathering. Deep cracks form.

thin, dry soil

scarp

cave

millstone grit and shale

limestone

clay

Water may react with chemicals in rock

There are other chemical reactions between water and rocks, which cause weathering. One of these is **hydration**. This happens when minerals in the rock combine with water to form new minerals. For example, haematite, Fe_2O_3, combines with water to form limonite, $Fe_2O_3.H_2O$. If the new minerals take up more space than the old ones, they force the rock apart as their crystals grow in the wet cracks. Sometimes only small flakes of rock fall off, but sometimes the whole outer skin of the rock can break away. This is called **exfoliation**. It produces rounded boulders, or huge, dome shaped outcrops of rock.

Hydrolysis is also caused by reactions between water and minerals, but it is not the same as hydration. In hydrolysis, water combines with minerals in the rock to form completely new substances. For example, water combines with potassium feldspar in granite to form **kaolin**, china clay. This has happened on a large scale in Cornwall, where china clay is mined. Hydrolysis of rock minerals also produces many other types of clays.

Oxidation of minerals also contributes to the weathering of rocks. Iron silicates can combine with oxygen to form iron oxide. This is a type of rusting, and leads to a red-brown stain on the surface of the rock.

— E X T E N S I O N —

Fig. 52.4   Limestone often forms this sort ▶ of landscape, called karst. The ridges are called clints, and the clefts are called grykes. The grykes are caused by weathering, as water penetrates the limestone and dissolves it away. Although the landscape looks barren, many plants live in the shelter of the grykes. This photograph was taken in County Clare, Ireland.

Living things can break rocks apart

Tree roots may push down through cracks in rocks. As they grow, they force the rocks apart. Even small plants can do this on the surfaces of rocks.

Burrowing animals such as earthworms and rabbits also play a small part in the breakdown of rocks into smaller particles.

Fig. 52.5   Weathering decomposes granite into kaolin or china clay. China clay is mined in several places in Cornwall, including here at St Austell.

53 SOIL

Soil is built up by the action of living things on weathered rock.

Soil is formed from rock debris and organic remains

The weathering of rocks produces small particles of various minerals. These include sand and clay. The relative amounts of sand and clay depend on the rock from which the particles were formed – the parent rock.

Particles of weathered rock – or even solid rock – can be colonised by **lichens**. The lichens absorb minerals from the rock, and grow slowly. Other small organisms – bacteria, small animals and small plants such as mosses – can begin to colonise the rock. Over time, they die. Bacteria and fungi begin to break down their remains. Rotted and partly-rotted organic material called **humus** is produced, which mixes with the rock particles.

The mixture of rock particles and humus is called **soil**. We are so used to it that we take it for granted. But a deep layer of soil takes hundreds of years to form. We need to take great care of it.

Fig. 53.1   A lichen is made up of two organisms, an alga and a fungus, living in very close association. The alga photosynthesises, and the fungus obtains minerals dissolved out of the rock. Together, they are able to grow where few other plants could survive.

Soil components

All soils have six main components. Weathered rock fragments, or **mineral particles,** make up the bulk of most soils. Coating these particles is a thin layer of dark brown, sticky **humus. Living organisms**, such as bacteria, plant roots and earthworms, are also an important part of any soil.

In between the mineral particles are **air spaces**. These are very important because oxygen from the air is used in respiration by most soil organisms. Nitrogen is also used by some bacteria, to make nitrogen compounds which plants need for growth.

The spaces between the particles also contain **water**. Without water, plants are unable to grow. Dissolved in the water are various **ions**. Some of these come from the mineral particles. Others, such as nitrates, are produced as bacteria and fungi break down the bodies of plants and animals. On agricultural land, ions needed by plants are added to the soil as fertilisers.

Fig. 53.2

Clay particles are small and plate-like. They pack tightly together, leaving little room for air.

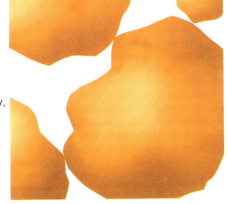

Sand particles are large and irregularly shaped. They cannot pack tightly, so there are large air spaces between them.

Fig. 53.3   Sandy soils are easy to cultivate, but have a tendency to dry out easily, or to blow away if left bare.

Different sizes of mineral particles produce different soils

The sizes of the mineral particles in a soil are a very important influence on its properties. **Sand** particles tend to be quite large. They cannot fit closely together, so there are large spaces between them. These spaces leave lots of room for air. Water can drain quickly through them. So sandy soils tend to be light, well aerated, and rather dry.

Clay particles are much smaller. They pack tightly together, leaving little room for air. Water is held in the tiny gaps between the particles, by capillarity. So clay soils tend to be heavy with little air, and hold more water than sandy soils.

Clay soils contain more ions than sandy soils. There are two reasons for this. Firstly, the minerals making up clay particles are charged. They attract ions in the soil water, and hold them on the surfaces of the particles. Secondly, water runs slowly through a clay soil, but quickly through a sandy soil. As water drains through a sandy soil, it carries dissolved ions with it. They soak downwards through the soil. This is called **leaching**.

Humus provides a good soil for plant growth

A good soil contains plenty of humus. Humus helps to hold water in the soil. Even a sandy soil can hold water, if there is plenty of humus in and around the sand particles.

Humus, like clay particles, attracts and holds ions. So humus helps to stop nutrients in the soil from being washed away. Humus increases the fertility of the soil.

Humus is also very good for clay soils. The humus sticks the small clay particles together into clumps, or **crumbs**. The crumbs are larger than the individual clay particles, so the soil has larger air spaces, is easier and lighter to work, and is better drained.

Soils which form in dry environments contain little humus. This is because not many plants can grow in them, so there is very little organic material for bacteria to rot down. Moist, tropical climates also tend to produce soils with little humus. This is because bacteria rapidly digest and oxidise the humus. The most humus-rich soils form in moist, temperate countries, such as Britain.

Fig. 53.4 Clay soils are heavy and sticky when wet, but when they dry out they become hard and cracked.

Fig. 53.5 You can make a rough estimate of the proportions of different sizes of particles in a soil by mixing a sample with water in a gas jar and allowing the particles to settle out. Large sand and gravel particles settle to the bottom, with smaller particles above. Tiny clay particles remain suspended in the water.

INVESTIGATION 53.1
Measuring the percentage of water and humus in a soil sample

1 Mass an evaporating dish. Fill it roughly $2/3$ full with a sample of soil. Mass it again. Subtract the mass of the empty dish from the mass of the dish of soil, to find the mass of the soil.

2 Put your dish of soil into a warm oven. A temperature of around 70 °C works well. You must not get the soil too hot, as you do not want to burn the humus at this stage. You just want to evaporate all the water off the soil.

3 After a day or so, mass the dish of soil again. It should have lost mass, because water will have been lost from the soil. (How can you tell if *all* the water has gone?) If you subtract this new mass from the original mass, you can find out how much water has been driven off from the soil while it was in the oven.

4 Now heat your dry soil sample to a very high temperature (about 400 °C). You may have an oven which is hot enough to do this. If not, you can burn your soil sample with a roaring Bunsen flame. Get all the soil really hot, to burn off all the humus. Then let it cool, and mass it. (As with the water you need to make sure that *all* the humus has gone.) Subtract this new mass from the mass before you burnt the soil sample, to find out how much humus there was.

5 Now calculate the percentage of water, and the percentage of humus, in your sample of soil. For water, use the formula:

$$\frac{\text{mass of water driven off in the warm oven}}{\text{original mass of soil}} \times 100$$

Calculate the percentage of humus using the formula:

$$\frac{\text{mass of humus burnt off}}{\text{original mass of soil}} \times 100$$

Questions
1 Why does the soil lose mass in the warm oven?
2 Why must the oven not be too hot?
3 Why does the soil lose mass when it is heated strongly?
4 Why must *dry* soil be used to find the amount of humus in the soil?

INVESTIGATION 53.2
The effect of lime on clay

1 Shake a small amount of clay with water, to make a suspension.

2 Add a little powdered lime to the clay, stir, then watch to see what happens. Record all your observations.

Questions
1 What is lime?
2 What did the lime do to the clay particles?
3 Farmers and gardeners often add lime to clay soil. Can you suggest how the lime improves the properties of the soil?
4 Lime is also added to acid soils. Explain what effect it would have.

Questions
1 Make a table to compare the following features of a sandy and a clay soil: particle size, ease of working, aeration, drainage, moisture holding ability, amount of mineral ions.
2 Find out about:
  a the uses which plants make of the water they obtain from soil.
  b the part played by soil bacteria in the carbon and nitrogen cycles.
  c the most important ions which plants obtain from the soil, and why they need them.

54 EROSION

Weathered rock and soil can be carried away by wind, water and ice.

Soil moves down hillsides

Soil on a hillside is pulled downwards by **gravity**. Other forces act against this downwards force. The main one is **friction** between the mineral particles in the soil. Plant roots help to increase the frictional force. Another force holding the soil particles in place is **cohesion** between water molecules and the molecules in the soil particles.

If the forces balance, the soil stays where it is. But if they do not, the soil moves down the hillside. The soil **erodes**. This is less likely to happen if there is a good covering of vegetation. The plant roots help to hold the soil particles in position.

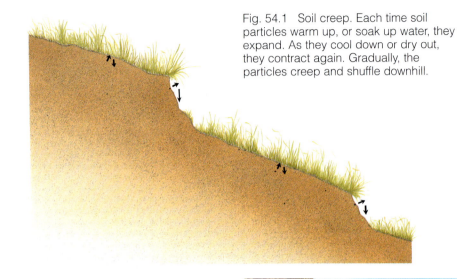

Fig. 54.1 Soil creep. Each time soil particles warm up, or soak up water, they expand. As they cool down or dry out, they contract again. Gradually, the particles creep and shuffle downhill.

Fig. 54.2 After the forest was cut down, rain water running over the surface of the ground washed away soil and formed this deep gully, near Kuala Lumpur in Malaysia.

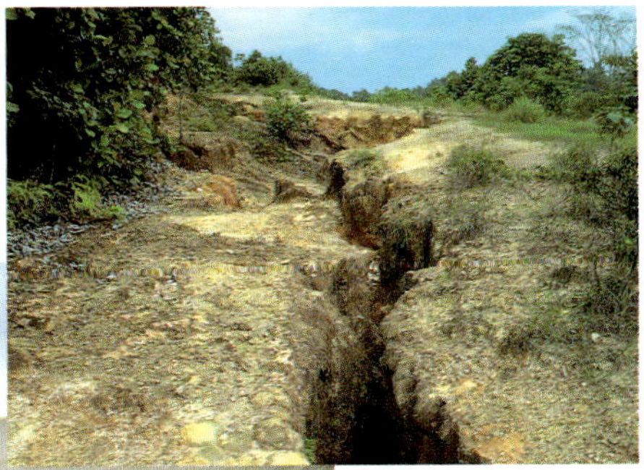

Soil creep is caused by alternate expansion and contraction

Soil particles can expand and contract. They expand if they get warmer, if they soak up rain, or if water inside them freezes. They contract if they cool down, or if they dry out. Small plant roots can also cause expansion of soil particles.

If a soil particle is on a slope, it tends to roll very slightly downhill when it expands. The same thing happens when it contracts. Very slowly, the soil particles move down the hillside. This is called **soil creep**.

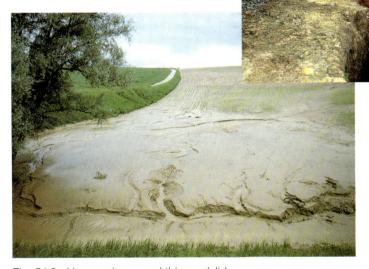

Fig. 54.3 Heavy rain caused this mudslide, which has eroded a huge amount of soil from the field. If there had been a good cover of crops on the field, the mudslide would probably not have happened.

Bare hillsides can be badly eroded by rain

When rain falls on a hillside covered with grass and trees, it will cause little if any erosion. The vegetation absorbs the impact of the rain on the soil. The rain soaks gently into the soil. Much of it is absorbed by the plants.

But on a bare hillside, it is a different story. The force of the raindrops landing on the soil splashes soil particles up into the air. They fall back to earth lower down the hillside. The soil particles gradually move down the hill. In some parts of the world this kind of erosion can carry away a depth of up to 150 cm of soil each year.

Heavy rain on bare slopes can cause landslides

Heavy rain falling on a bare slope soaks into the unprotected soil. As the soil soaks up the water, it becomes heavier. The force of gravity on it is greater. Moreover, the frictional force between the soil particles is reduced, because the water acts as a lubricant.

The upwards and downwards forces on the soil are no longer balanced. The soil moves swiftly down the hillside. Huge **mudslides** and **landslides** can happen in this way.

The sea and rivers carry away rocks and soil

Salt spray from waves splashing onto the shore can cause chemical weathering of rocks. This can cause the rocks to crumble, and fall into the sea. Cliffs can be eroded in this way.

Big waves hitting the shore can speed up this process. Large air pressures are created between the waves and the cliffs as the waves approach. These pressures can act on cracks in the cliffs, causing parts of the cliffs to break away.

Waves, and fast-moving water in rivers, can directly erode rocks by the force with which they hit them. This is called **hydraulic action**. Waves and rivers may also carry sand and gravel, which grates against rocks and erodes them. This is called **abrasion**. Hydraulic action and abrasion produce small rock particles, which are carried away by the sea or the river.

Fig. 54.4 The sea has eroded these cliffs at Easton Bavents in Suffolk. The erosion here is very rapid, and the house nearest the edge of the cliff has been demolished since this photograph was taken.

Fig. 54.5 Most of the time, rivers cause little erosion, but when in flood the effects can be devastating. This damage was caused by a flooded river in the Tyrol, Austria.

Glaciers erode mountains

Snow collects in hollows on mountains. As the weight of snow builds up, it becomes compacted and forms ice. If large amounts of ice are formed, the weight of the ice causes it to move slowly downhill. A **glacier** is formed.

The moving ice collects particles of rock debris. These are carried along with the ice, scratching away at the rock over which the ice flows. So the glacier erodes the rock over which it moves.

Light, dry soils can be eroded by wind

Soil can be eroded even if it is not on a slope. This is especially true of light, dry sandy soils. If they are unprotected by vegetation, these soils can be blown long distances by the wind.

Fig. 54.6 Wind has eroded these sandstone towers in Paracas National Park, Peru. The towers are between 15 and 25 cm high.

55 DEPOSITION

Material eroded by wind, ice or water is eventually dropped or deposited.

Rivers deposit sediment when water slows down

The amount of material which a river can carry in suspension depends on how fast it is flowing. The faster it flows, the more material it can carry and the bigger the particles which can be carried in suspension in the water. Figure 55.1 shows how the velocity of flow of a river affects the size of material which can be carried by it.

When the water in a river slows down, it will deposit some of the material which it is carrying. This will happen when the river flows into a lake or reservoir, or into an estuary, or when the river broadens out over a wide valley floor. The larger particles are deposited first. Wide valley floors, such as the valley of the Thames, often have deep deposits of sand and gravel, dropped by the river as it slows down. Smaller particles, such as silt, are carried much further before they are dropped. They are often carried all the way to the sea.

Flooding increases erosion and deposition

After very heavy rain, a lot of extra water flows into a river. Rivers flow faster than usual, and carry a lot of extra sediment. There may be so much water in a river that it floods over its banks. As the water escapes from its normal channel, it may spread over a wide area on either side. As this flood water slows down, it drops its load.

Figure 55.3 shows a section through a flooded river. The further the flood water is from the river channel, the slower it flows. So the water flowing in the channel can carry larger particles than the water flowing more slowly at the edge of the flooded area. As the flood water subsides, large particles, such as sand and gravel, are deposited near to the river channel. Smaller particles, such as mud and fine silt, are deposited further away.

Fig. 55.1

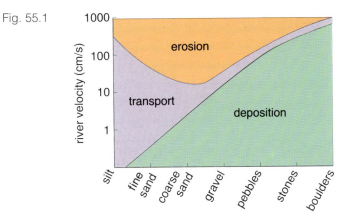

increasing size of particles

Fig. 55.2   The estuary of the River Blyth in Suffolk contains deep deposits of mud dropped by the river as it slows on reaching the sea. Although not everyone finds muddy estuaries beautiful, the mud contains a lot of nutrients, and burrowing animals such as worms live in it. So mud is an important feeding ground for many kinds of birds.

Fig. 55.3   A river flows through an area of low-lying land called a **flood plain**. After heavy rain, the river may overflow its channel. Water spreads out over the flood plain and deposits sediment.

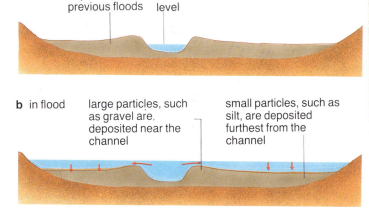

a normal   sand and gravel deposits from previous floods   normal water level

b in flood   large particles, such as gravel are, deposited near the channel   small particles, such as silt, are deposited furthest from the channel

Fig. 55.4   Loess is deep and fertile, but it is also very easily eroded by wind and water. These Chinese people are planting poplar trees to try to prevent erosion. The terraces on the hillsides in the background are also intended to reduce erosion, and allow people to farm even on very steep slopes.

134

The sea deposits eroded material which forms beaches

Rock material eroded by the sea may be carried long distances before being deposited. Some of it is deposited on the sea bed. But quite a lot is deposited on the shore, where it forms **beaches**.

The loose grains which form sandy beaches are easily moved away again by waves. Waves may carry sand along a beach. This is called **longshore drift**. Towns which rely on their beaches to attract tourists often build **groynes** along the beach, to stop the sand from being carried away.

Fig. 55.5 Longshore drift has been checked at Southwold in Suffolk by building groynes along the beach. In which direction is the sea moving the sand?

Wind drops eroded material forming dunes

Wind can carry away particles of loose, dry soil and weathered rock. These particles are eventually deposited as the energy of the wind decreases. They may form **dunes**.

Millions of years ago, north-westerly winds blew across the Gobi Desert in central Asia. They carried away the fine, dry soil from this area. The soil was deposited in what is now north-east China. Huge amounts built up. This type of soil is called **loess**, and is deep and fertile.

However, much of the natural vegetation has now been removed from the loess by the millions of people who live and farm in north-east China. Now the loess is being eroded again, this time by water. The water washes the loess into rivers. The Yellow River got its name because of the huge amounts of sediment which it carries towards the sea. This erosion is very damaging. It is depriving the farmers of the loess region of their soil and it is causing dangerous flooding in the land around the Yellow River. An intensive program of tree planting and terracing in the loess region is now being carried out, in an attempt to halt the erosion.

INVESTIGATION 55.1

Erosion and deposition by waves

Design and carry out an investigation into the erosion and deposition of sand or gravel by wave action. Use a shallow tray (such as a seed tray or laboratory storage tray), and make a diagonal beach along one long edge. Half fill the tray with water. Make waves by moving a piece of card or a ruler up and down, or *very gently* backwards and forwards.

You could investigate:
- from which areas of the beach the sand is eroded, and where the eroded sand is deposited.
- in which direction the sand moves relative to the direction of the waves and the angle at which they hit the beach.
- the effects of different wave amplitudes or frequencies.
- whether erosion and deposition are the same for the different sizes of particles in the beach.
- whether putting groynes in the beach reduces erosion.

Questions

1 The River Tamar runs along the boundary between Cornwall and Devon. The graph below shows data collected between 1984 and 1986, for the average flow rate of the river, and the amount of suspended sediment in the water.

   a What was the maximum flow rate over this period?

   b What was the maximum amount of suspended sediment over this period?

   c What relationship can you see between the flow rate and the amount of suspended sediment?

   d What is likely to happen to the large amounts of suspended sediment which the River Tamar sometimes carries?

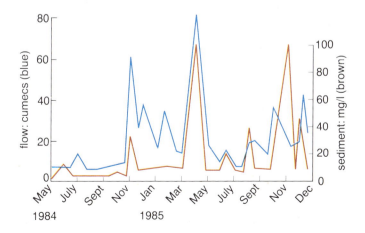

56 THE ICE AGES

An ice age is a long period when the Earth's climate cools down so much that the polar ice sheets advance towards the equator. The last one ended about 10 000 years ago.

The cause of ice ages is not certain

We can find out about the climate of the past by looking at rocks, and the fossils they contain. The kinds of fossils found give clues about the temperature in which the animals and plants might have lived. Another way of working out what the temperature might have been in the past is to analyse the oxygen isotopes in fossil shells. The proportion of the different isotopes which are used in the formation of a shell depends on the temperature of the water in which the shelled animal is living.

Investigations like these show that the Earth's climate has changed considerably over the past few million years. There have been several very long periods when the climate was much cooler than it is today. The last ice age was in the Pleistocene. It began about 2 000 000 years ago, and ended 10 000 years ago. The Earth's climate became so cold that the polar ice sheets spread northwards and southwards. Figure 56.1 shows the extent of the ice sheets about 20 000 years ago.

No-one is quite sure why ice ages happen. Some possible causes include:

- the Earth's orbit around the Sun changing over very long periods of time, sometimes bringing the Earth a little closer to the Sun, and sometimes taking it further away.
- storms on the Sun's surface, which appear as sunspots, reducing the amount of radiation being emitted from the Sun, so cooling the Earth.
- dust clouds in space drifting between the Earth and the Sun.
- the composition of the Earth's atmosphere changing for some reason. For example, the increase in carbon dioxide in the atmosphere may cause the Earth to warm up. Or large amounts of dust from an erupting volcano could reflect the Sun's rays, preventing them from reaching the Earth's surface and so cooling it down.

Probably, each ice age had its own set of causes. It is very difficult to predict if and when the next one might happen. Some people have calculated that it will begin 23 000 years from now. Global warming might delay or even prevent another ice age, but we just don't know.

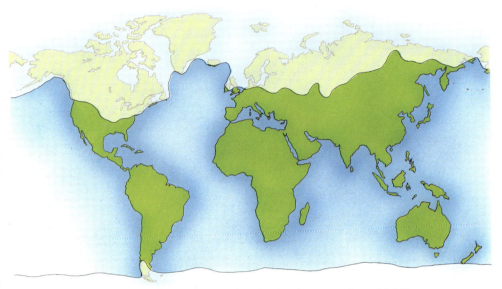

Fig. 56.1   The extent of the ice sheets during the last ice age, about 20 000 years ago.

Fig. 56.2   During the last ice age, most of Britain was under an ice-cap. The seas near the edge of the ice-cap contained pack ice and icebergs.

The last ice age affected the landscape of Britain

As you will see from Figure 56.2, the ice sheet in the last ice age covered most of Britain. Snow collected in very deep layers. The snow became compacted, forming ice. Rivers of ice, or **glaciers**, formed. The flow of the glaciers, eroding rock material from one place and depositing it in another, changed the landscape. When the last ice age ended and the ice melted, it left behind many features which are still visible today.

As a glacier flows slowly downhill, it drags rock particles down with it. The pieces of rock scratch and scour along the surface over which the glacier flows, slowly hollowing out a deep valley. Glaciers can erode a surface up to twenty times faster than a river can. The lakes in the Lake District in Britain are glacial valleys, now filled with water.

The load of rocks and soil which a glacier collects and carries along is called its **moraine**. When the glacier begins to melt, this material is left behind. It is a jumbled mixture of all different sizes and shapes of rock pieces, large boulders, and soil. It is called **till**. Till is deposited all along the glacier's path, leaving deep layers which cover the ground. Heaps of till also form along the edges of the glacier, and at its snout. These piles of jumbled rock and soil are also called **moraines**, and can be seen in many mountain valleys in Britain.

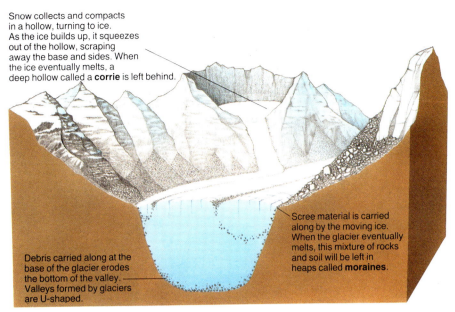

Snow collects and compacts in a hollow, turning to ice. As the ice builds up, it squeezes out of the hollow, scraping away the base and sides. When the ice eventually melts, a deep hollow called a **corrie** is left behind.

Scree material is carried along by the moving ice. When the glacier eventually melts, this mixture of rocks and soil will be left in heaps called **moraines**.

Debris carried along at the base of the glacier erodes the bottom of the valley. Valleys formed by glaciers are U-shaped.

Fig. 56.3 Glaciers erode and deposit rock particles.

Fig. 56.4 A glacier in Greenland. This glacier is only carrying a small amount of rock and soil along with it, so it looks clean and bright. Some glaciers are so full of rock particles that they look brown. If this glacier melted, the valley in which it has been flowing would be seen to be a U-shape, carved out by the moving ice.

DID YOU KNOW?

At one time, during the Pleistocene Ice Age, ice sheets covered 30% of the Earth's surface.

Questions

1 The chart on the right shows the climate at different latitudes in the Northern Hemisphere 18 000 years ago.

a Use your atlas to draw a similar chart for the present day.

b Briefly discuss the differences between the positions of the climatic zones in the last ice age and today.

c What would it have been like in Britain 18 000 years ago? What parts of the world today have a climate like this?

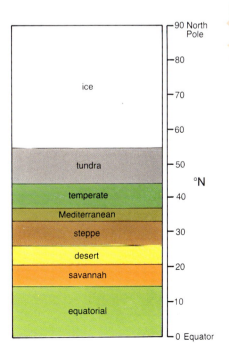

ice	90 North Pole
	80
	70
	60
tundra	50
temperate	°N 40
Mediterranean	
steppe	30
desert	20
savannah	
equatorial	10
	0 Equator

Fig. 56.5 Glen Torridon in Wester Ross, Scotland. The hillocks are heaps of boulders, rocks and soil left behind when a glacier melted about 10 000 years ago. This debris is called moraine. The mountains behind the hillocks were scraped clean by the glacier.

THE EVOLUTION OF THE ATMOSPHERE

When the Earth was first formed, the composition of the atmosphere was very different from today.

The Earth's first atmosphere contained no oxygen

It is thought that the Sun and the nine planets, which together make up the solar system, were formed about four and a half thousand million years ago. The Earth would have been very hot then, and some scientists think that it would all have been molten and in a liquid state. The solid crust gradually formed as the young Earth cooled.

During this time, gases were released from the Earth. They would have bubbled out from the surface. Later, as the crust formed, many volcanoes developed. When they erupted, they released huge quantities of gases.

Some of these gases, such as hydrogen and helium, would have been too light to have been held close to the Earth by its gravity. These gases would have escaped into space. Other gases stayed close to the Earth, forming its atmosphere. These included water vapour, carbon monoxide, carbon dioxide and small amounts of ammonia and methane. There would have been no oxygen.

Fig. 57.1 Volcanoes released enormous amounts of gases into the early Earth's atmosphere. Most of the gas released from a volcano is water vapour and carbon dioxide, but sulphur dioxide, hydrogen chloride, hydrogen sulphide, methane and hydrogen may also be present.

The amount of carbon dioxide in the atmosphere decreased with time

As the Earth gradually cooled, much of the water vapour in the atmosphere condensed to form liquid water. This is how the oceans were formed. Some of the gases in the Earth's atmosphere dissolved in the water in the newly-formed oceans. Carbon dioxide, for example, will dissolve in water and then react to form carbonates. Some of these carbonates formed sediments at the bottom of the oceans. This reduced the amount of carbon dioxide in the atmosphere.

Life first evolved on Earth between about three and a half thousand million, and four thousand million, years ago. Fossils in rocks formed two and a half thousand million years ago show that by then bacteria which photosynthesised in a similar way to modern green plants had become abundant. Later, other photosynthesising organisms evolved. These included tiny green protoctists in the sea, and also larger plants growing on land. All of these photosynthesising organisms would have taken large amounts of carbon dioxide from the atmosphere.

Where did this carbon dioxide go? It became part of the bodies of the photosynthesising organisms, and then part of the bodies of animals that ate them. Some of this carbon dioxide would have gone back into the atmosphere as the organisms respired. Some more of this carbon dioxide would have returned to the atmosphere after the organisms died, when decomposers of the bodies of organisms respired. But the bodies of many plants were not broken down by decomposers. Instead, they became buried in swampy ground, and eventually turned into fossil fuels, which contain a lot of carbon.

Some of the carbon taken in by the plants became calcium carbonate in shells of protoctists and animals. These shells dropped to the bottom of the sea, where they formed limestone rocks.

All of these processes reduced the concentration of carbon dioxide in the atmosphere. Today, only about 0.4% of the atmosphere is carbon dioxide.

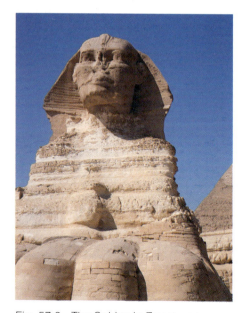

Fig. 57.2 The Sphinx in Egypt was carved out of limestone rock. The limestone is calcium carbonate, formed from the shells of countless millions of tiny marine organisms.

Photosynthesis added oxygen to the atmosphere

When living organisms began to photosynthesise, they changed the Earth's atmosphere for ever. For the first time, large amounts of oxygen were released into the atmosphere.

Before this happened, all the living organisms that had evolved were anaerobic. This means that they lived without oxygen. When oxygen appeared in the atmosphere, some organisms evolved the ability to respire aerobically. If there had been no photosynthesis, there would still be no oxygen in the atmosphere, and animals would never have evolved.

The oxygen had other effects, too. It reacted with chemicals in rocks that were in contact with the atmosphere. For example, iron sulphide, FeS_2, would have reacted to form iron oxide. Today, iron sulphide is never found on the surface of the Earth. But it can be found in sedimentary rocks buried deep in the Earth, which formed more than two thousand million years ago, before there was much oxygen in the Earth's atmosphere.

In the very high parts of the atmosphere, about 12 to 24 km above the Earth's surface, ultraviolet radiation from the Sun causes some of the oxygen, O_2, to react and form ozone, O_3. This layer of ozone absorbs ultraviolet light and reduces the amount that reaches the Earth's surface. Ultraviolet light can damage the DNA in living organisms, causing mutations. The ozone layer helps to prevent this from happening.

Fig. 57.3 The gold-coloured crystals are a form of iron sulphide called 'fool's gold'. If iron sulphide is present in a rock, it indicates that it was formed under conditions where no oxygen was present.

The amount of nitrogen in the atmosphere has increased

Today, most of the Earth's atmosphere consists of nitrogen gas, N_2. The early atmosphere would not have contained much nitrogen gas, but it would have contained ammonia, NH_3. Some of the early organisms to evolve were probably nitrifying bacteria. They would have converted ammonia into nitrates, so removing ammonia from the atmosphere.

Ammonia would also have been removed when oxygen began to be produced by photosynthesising organisms, because ammonia reacts with the oxygen to produce nitrogen gas. Another reason for the increase in nitrogen gas in the atmosphere is the conversion of nitrates into nitrogen by denitrifying bacteria.

The atmosphere has not changed much in the last 200 million years

The changes described above all took place a long time ago. For the last 200 million years, the composition of the Earth's atmosphere has stayed relatively constant. Plants continue to release oxygen and take in carbon dioxide for photosynthesis, but approximately equal amounts are taken in and given out by living organisms for respiration. Nitrogen gas is still produced by denitrifying bacteria, but approximately equal amounts are taken from the atmosphere by nitrogen fixing bacteria, in the manufacture of ammonia in the Haber process, and as nitrogen reacts with oxygen when lightning flashes through the air.

But, in the last two thousand years or so, the human population on Earth has greatly increased. Our activities are causing changes in the Earth's atmosphere. We are burning large amounts of fossil fuels, which may cause an increase in the amount of carbon dioxide in the atmosphere. We have released chemicals called CFCs into the air, which react with ozone and have reduced the amount of it in the high levels of the atmosphere.

Questions

1 Copy and complete the table, to show the major differences between the Earth's atmosphere when it first formed, and the atmosphere today.

Gas	Was it present in the early atmosphere?	Is it present in today's atmosphere?	Is there more or less of it now than when the atmosphere first formed?
water vapour	yes	yes	probably less
nitrogen			
oxygen			
carbon monoxide			
carbon dioxide			
ammonia			
methane			

2 Many different processes have contributed to changes in the concentration of gases in the Earth's atmosphere in the last four thousand million years. Match each process listed below with one or more changes. You should use each change at least once. One process contributed to all four changes!

processes photosynthesis
      formation of limestone rocks beneath the sea
      formation of fossil fuels
      evolution of nitrifying bacteria

changes reduction in the amount of carbon dioxide in the atmosphere
      appearance of oxygen in the atmosphere
      reduction in the amount of ammonia in the atmosphere
      production of the ozone layer

1 Figure 43.2 on page 108 shows the main plate boundaries on the Earth's crust. The arrows show the direction of movement of the plates.

a Make a copy of this map.

b Name the countries or continents labelled A to G.

c Describe the position of one plate boundary where:

  i two plates are moving apart.

  ii two plates are pushing into one another.

  iii two plates are sliding past one another.

You could mark these boundaries on your map, using different colours for each kind.

d Use your atlas to mark the position of the following mountain ranges:

  i the Andes

  ii the Rocky Mountains

  iii the Alps

  iv the Himalayas.

What do you notice about the positions of these mountains? Can you explain your observations?

e Use your atlas to mark the position of San Francisco. Can you explain why earthquakes happen so frequently in this area?

f Use reference books, including your atlas, to find the names of at least ten volcanoes. Mark them on your map. Can you explain why they are found in these areas?

2 Read the passage on the right, and then answer the questions which follow.

a What is an SSSI?

b Why has this area at Walton-on-the-Naze been designated an SSSI?

c What is causing the erosion at Walton-on-the-Naze?

Erosion of an SSSI

Walton-on-the-Naze is on the East Anglian coast. The map shows a part of this area which has been designated a site of special scientific interest. The reason for this designation is the geology of the area. The lower parts of the cliffs are formed from London clay, which was laid down about 60 000 000 years ago. Above the clay lie soft red sandstones, deposited about 1 000 000 years ago. The clay deposits are rich in fossil bird bones, which are found in large numbers and are very well preserved. The Naze is one of the most important sites of bird fossils in the world. The red sandstones are important, too. They are made up of shelly sands, and red-brown sands, deposited by the sea. They give geologists information about what was happening in Britain just before the last ice age.

The sea constantly erodes the cliffs at Walton-on-the-Naze. Geologists want the erosion to continue. It helps to keep the cliff face clear and well defined, so that they can study it. It stops vegetation from growing on the cliff face. As the erosion continues, it exposes new fossils.

But people who live in the area want the erosion to be prevented. They are worried about the loss of land as the sea washes it away. They want to be able to walk along the cliff paths close to the sea, without having to worry that they will fall away. A plan has been put forward to build breakwaters along the base of the cliff, and a sloping wall about 5 m high, to reduce erosion. This would cost between £1 000 000 and £3 000 000.

d Why do geologists want the erosion to continue?

e Why do local people want the erosion to stop?

f How would the breakwaters and the sloping wall help to prevent erosion?

(Source – Nature Conservancy Council, 1988)

3 Read the following passage, and then answer the questions which follow.

Rainstorms and soil erosion

Raindrops falling onto ground with poor vegetation cover are a major cause of erosion. In tropical regions, where much of the rain falls in sudden, heavy bursts, this type of erosion is particularly important.

The size of the raindrops which hit the ground has a big effect on the amount of damage they do. In Britain, raindrops tend to be quite small. Even in heavy rain, they are rarely above 2 mm in diameter. Scientists can measure the diameter of raindrops by catching a sample of them on a tray of freshly-sieved flour. The size of the pellet which forms when a drop hits the flour is related to the size of the raindrop.

Raindrops in the tropics tend to be much larger. In one study in Nigeria, it was found that most raindrops were between 2 and 3.5 mm in diameter. However, some had diameters as great as 5.5 mm.

The bigger the raindrop, the more energy it has when it hits the ground. So large raindrops tend to cause more erosion than small raindrops.

The intensity of rainfall also affects the rate of soil erosion. In the study in Nigeria, 23.3 cm of rain fell in one hour during a particularly violent storm. In Britain, rainfall hardly ever comes anywhere near this figure. It is estimated that the energy of rain as it hits the ground is about 37 J/m^2 for each millimetre of rainfall.

Not surprisingly, storms such as the one in Nigeria can wash away huge quantities of soil. On a one or two percent slope, soil loss can be as much as 55 tonnes per hectare every year. In Madagascar, erosion of up to 250 tonnes per hectare has been recorded.

a It is possible to produce water drops of known sizes. Assuming that you can do this, explain exactly how you would use a tray of sieved flour to measure the average size of raindrops in a shower of rain.

b Give two reasons why the energy with which rain hits the ground in some tropical regions is greater than in Britain.

c Discuss the ways in which erosion caused by rainfall can harm the environment.

d Discuss the possible ways of reducing this type of soil erosion, and the reasons why countries such as Madagascar cannot immediately put these methods into practice.

cliffs and higher ground

beach

proposed sloping wall

proposed breakwaters

boundary of SSSI

RESOURCES

The Earth contains huge numbers of resources which we use to release energy or to make objects. Some of these resources are running out.

Resources may be renewable or non-renewable

All of our many manufactured articles are made from the Earth's natural resources. We obtain all our materials, and many of the resources which we use to provide energy, from the Earth.

Some resources are being constantly produced on the Earth. They are called **renewable** resources. Others are not being replaced. They are called **non-renewable** resources.

Renewable resources include solar energy and plants

Many things which we obtain from plants and animals can be considered renewable resources. Such resources include cotton, wool and leather for clothing, and wood for timber and paper.

Cotton is obtained from plants which are grown just so that cotton can be obtained from them. When the plants have been used, new ones are grown in their place. The supply of cotton will never run out. But much of the wood which is used for building, or making paper, comes from trees which are cut down and not replanted. If we do not take care, this resource may be in short supply in the future.

Some sources of energy are renewable resources. Solar power, hydroelectric power and wind power will never run out.

Before the industrial revolution, renewable resources were much more important than non-renewable ones. People wore clothing made from natural fibres, such as cotton, linen, wool or animal skins. Water-mills, windmills and firewood were the chief sources of energy.

Non-renewable resources include fossil fuels and minerals

The Earth contains many resources which are not being replaced. These include **minerals** buried in the Earth's crust. Some of these minerals are important sources of metals. They are called **ores**. As we dig them up, the supply in the ground decreases.

Fossil fuels such as oil, coal and natural gas are very important sources of energy and chemicals in our modern industrial society. They were produced many millions of years ago, and they are not being replaced.

With many of our non-renewable resources, however, there is no real problem. **Silicon**, for example, is an element used in the manufacture of electronic chips. It can be obtained from sand. But the world's consumption of silicon is tiny compared with the amounts of sand available. Similarly, the extraction of **sodium chloride** from sea water will not make the sea any less salty!

But many non-renewable resources are in real danger of running out. The prices of many metals have gone up in recent years, as ore reserves become depleted. The rate at which we are using oil makes it a real possibility that it will be in very short supply during the next century.

Fig. 58.1 A cotton field in Louisiana, USA. The cotton fibres are part of the seed heads, or bolls, of the cotton plants.

Fig. 58.2 People in many parts of the world still use only renewable resources. This irrigation system uses mud, wood and leather as construction materials, and animal food as fuel.

Fig. 58.3 Oil is an important natural resource. Britain's largest oil field lies beneath the North Sea. This offshore production platform is called Brent Charlie, where drilling for, and extraction of, oil and natural gas take place.

Non-renewable resources must be used more efficiently

The problem of future shortages of some of our important resources must be faced. The answer to this problem will probably be a combination of four approaches.

- **Relying more heavily on renewable resources**. If a non-renewable resource becomes scarce, its price goes up. This can result in an alternative renewable resource becoming cheaper than the non-renewable resource, so the renewable resource will be used instead.

- **Increasing product lifetime**. We must try to make products made from non-renewable resources last longer. For example, most cars are made of steel, which contains large amounts of iron. The iron rusts. If every car could be made to last longer, then less iron would be needed by the motor industry.

- **Recycling**. Much of what we throw away could be reused. Glass, for example, can be crushed, remelted and reused. Glass made in this way requires much less energy than glass made from scratch. Many people already take their empty glass bottles to bottle banks. If more and more people were encouraged to do this, the amount of resources needed to make new glass could be greatly reduced.

- **Opening up new sources**. Improved technology is continually allowing resources which were previously unavailable to be extracted from the Earth.

Fig. 58.4 Quite a high proportion of glass in Britain is now recycled. It would be even better if the bottles themselves could be re-used.

Questions

1 a List the problems associated with renewable and non-renewable resources.

b Explain why our exploitation of non-renewable resources could be termed 'short-sighted'.

2 Glass and paper are two materials which can be recycled.

a What resource is conserved when glass is recyled?

b What resource is conserved when paper is recycled?

c Discuss the reasons for the relatively small amounts of glass and paper which are recycled in Britain.

d What can be done to increase the amounts of glass and paper which are recycled?

3 Figure 58.5 shows the estimated lifetimes of some important resources.

a Are the resources shown in this graph renewable or non-renewable resources?

b Which are fossil fuels?

c Which of these resources will run out first?

d Which has the greatest total reserves?

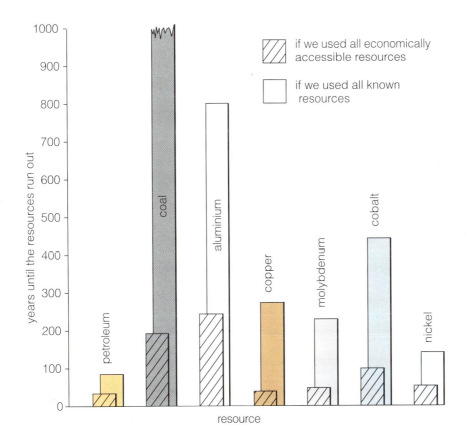

if we used all economically accessible resources

if we used all known resources

Fig. 58.5 Lifetimes of some important resources. These estimates assume that we will continue to use these resources at the same rate as we are doing now.

59 FOSSIL FUELS

Fossil fuels were formed from the remains of living organisms. They are important energy resources and also sources of many chemicals.

Coal, oil and natural gas are important energy sources

Amongst our most important non-renewable resources are the three fossil fuels – coal, oil (petroleum) and natural gas (methane). They are major energy sources. All three are used in the power stations which generate over half of Britain's electricity. All three are used for space heating in homes. Most of our road vehicles use oil products like petrol and diesel.

These three fossil fuels are also important sources of industrial chemicals. Coal is used to make coal tar. Coal tar is used to make nylon. Coal itself is used as a form of carbon for smelting iron. Petroleum is a source of a wide range of chemicals, including many plastics. Methane is the source of hydrogen used for making ammonia, which is used to make fertilisers.

Fossil fuels were formed millions of years ago

About 300 000 000 years ago, in the Carboniferous period, much of the Earth's surface was covered with dense forests and swamps. As trees and other plants died and fell, they did not rot away completely. The wet, airless mud made it difficult for decay bacteria to live. So the half-rotted trees gradually sank into the mud.

More and more dead vegetation and sediments collected on top of them. The half-rotted trees were crushed and flattened. They formed **coal**.

Oil formed during the same period. Oil formed under the seas. Tiny plants, animals and bacteria sank to the bottom of the sea when they died, mixing with the sand and silt on the sea bed. Their bodies did not rot away completely, but were crushed by the weight of sediments falling on top of them. Their remains became a mixture of carbon compounds – petroleum or crude oil.

Gas is found in the same rocks as coal and oil. It, too, was formed from the half-decayed remains of plants, animals and bacteria. Gas is often only found in small amounts, because it can easily leak away through tiny spaces in rocks.

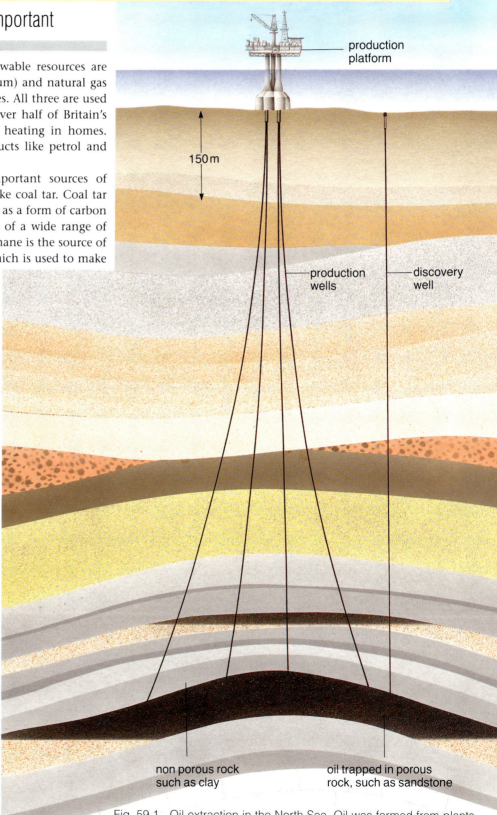

Fig. 59.1 Oil extraction in the North Sea. Oil was formed from plants and microorganisms which died millions of years ago. The oil can seep into rocks with tiny spaces in them (porous rocks). If there is a non-porous rock above, then the oil is trapped. Folded rocks form traps where oil may collect at the top of the fold.

Fig. 59.2 Coal was formed from trees and other plants. It sometimes contains fossils, such as this fern frond.

Coal is dug out of mines

We get coal by digging it out of the ground. Coal was formed in horizontal layers, called **seams**. In between the seams are layers of other rocks. A mine shaft is dug down through these other layers to reach a coal seam. Horizontal tunnels or galleries are then dug into the coal.

Sometimes the rocks will have been folded and twisted since they were formed. This can bring the coal close to the surface. The coal can then be dug out in an **open-cast mine**, rather like a quarry.

Rock movements or erosion by rivers may form a valley, cutting down through rock layers. A coal seam might then be exposed on the sides of the valley. Miners can dig straight into the valley side. This is called a **drift mine**.

Oil is pumped out of porous rocks

Oil does not form underground 'lakes'. Oil moves through porous rocks underground, until it gets trapped by non-porous rocks above it. It is held in the porous rocks like water in a giant sponge. To get at the oil, drills cut down through the non-porous rocks into the porous ones.

The oil may be under pressure, perhaps because there is gas trapped in the same place. If this happens, then the oil will surge up out of the drilled shaft. But often the oil has to be pumped out. Sometimes, a detergent solution is pumped into the oil-bearing rocks. The detergent dissolves the oil, and the mixture of oil and detergent is then pumped out.

Natural gas does not need to be pumped out of the ground. It will come out under its own pressure, once a shaft has been drilled down to it.

Fig. 59.3 An open-cast mine

Questions

1 a Why are coal, oil and gas called 'fossil fuels'?

b What are the two main uses of fossil fuels?

c Why are large amounts of fossil fuels no longer being formed?

d Where would you look to see if coal was still being formed?

e Where would you look to see if oil was still being formed?

f Coal is not explosive, but many mining disasters have involved explosions. What caused these explosions?

─ E X T E N S I O N ─

2 Britain's main oil field is under the North Sea. What problems does this give to the companies trying to extract the oil? Try to find out how some of these problems have been solved.

145

Coal is used to make coke, ammonia and coal tar

At the present time, well over half of the coal mined in Britain is burnt in power stations. But coal can also be used as a source of chemicals. In the past, a lot of coal was used in this way. It is now cheaper to obtain these chemicals from oil. But as oil and gas supplies begin to run out, we may soon need to turn to coal again.

Coal is heated in the absence of air. (If air was present, the coal would burn.) Gases and tarry products are driven off from the coal, leaving **coke**. Coke is an excellent smokeless fuel. The fumes from the heated coal are passed through water, where some components dissolve. One of these soluble components is **ammonia, NH$_3$**, from which fertilisers can be made. Other substances in the fumes form an oily layer on the surface of water. This layer is called **coal tar**. Coal tar contains many useful substances, including some which can be used for making plastics.

The rest of the fumes pass through the water unchanged. They are called **coal gas**. Coal gas is a good fuel.

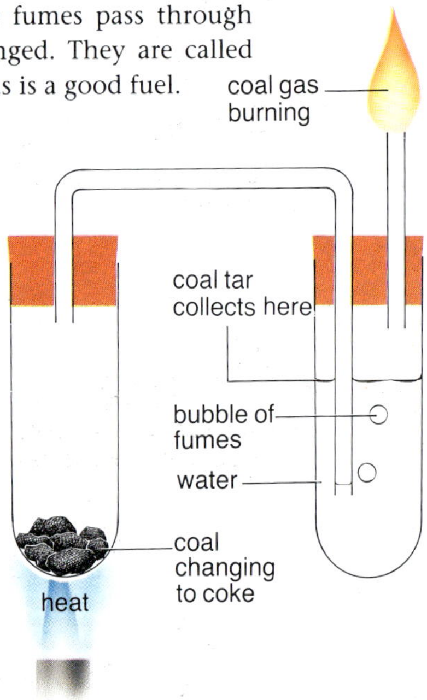

Fig. 60.1 A laboratory scale demonstration of coal as a natural resource.

coal gas burning

coal tar collects here

bubble of fumes

water

coal changing to coke

heat

Fig. 60.2 If the coke burning here is pure carbon, what are the products of this chemical reaction?

Natural gas is used to make ammonia

Natural gas is mostly composed of **methane, CH$_4$**. It can be used as fuel. Its other main use is as a source of **hydrogen.** Methane is mixed with steam, and passed over a nickel catalyst where a reaction takes place.

$$CH_4(g) + 2H_2O(g) \rightarrow CO_2(g) + 4H_2(g)$$

The carbon dioxide is removed. The hydrogen is mixed with nitrogen which has been extracted from the air. The hydrogen and nitrogen are reacted together to make **ammonia**. Methane, water and air have replaced coal as the world's main source of ammonia.

Oil is refined by fractional distillation

Oil is a mixture of so many valuable chemicals that it has been nick-named 'black gold'. Many of these chemicals are liquids, but others are solids and gases dissolved in the liquid. The mixture of chemicals is separated or **refined** using **fractional distillation**. This process works in the same way as the process for separating a mixture of alcohol and water. The different components separate because they have different boiling points.

Fig. 60.3 A crude oil fractionating column at Pulau Bukom, Singapore.

The untreated, or crude, oil is heated and injected into the bottom of a **fractionating tower**. The tower is very hot near its base, so most of the oil boils and the vapour rises up to the next level. The parts of the oil which do not boil include wax and tar. These parts, or **fractions**, of the oil are collected from the base of the tower.

The next level up, where the vapours go, is a little less hot than the base of the tower. So part of the vapour, including components such as lubricating oils, condenses here and can be collected. The rest of the oil, still in the form of vapour, rises up to the next level, which is a little less hot, and so on. The many levels of the fractionating tower separate the oil into its many fractions.

Questions

1 a Why will the exhaustion of oil and gas supplies affect our use of coal? What changes will take place?

b Why is coal rarely used now as a source of chemicals?

2 a What is the job of an oil refinery?

b What is the main device used in refining oil?

c What is meant by a crude oil 'fraction'?

d Which crude oil fraction is used to power:

   i ships?   **iii** jet aircraft?

   ii cars?   **iv** lorries?

e Which crude oil product may be found in a medicine cupboard?

---

3 a In the reaction in which hydrogen is obtained from methane, what does the nickel catalyst do?

b How could you remove the carbon dioxide from a mixture of carbon dioxide and hydrogen?

c Methane can also react with steam to give carbon monoxide and hydrogen. Write a balanced chemical equation for this reaction.

d How might nitrogen be extracted from the air?

— *E X T E N S I O N* —

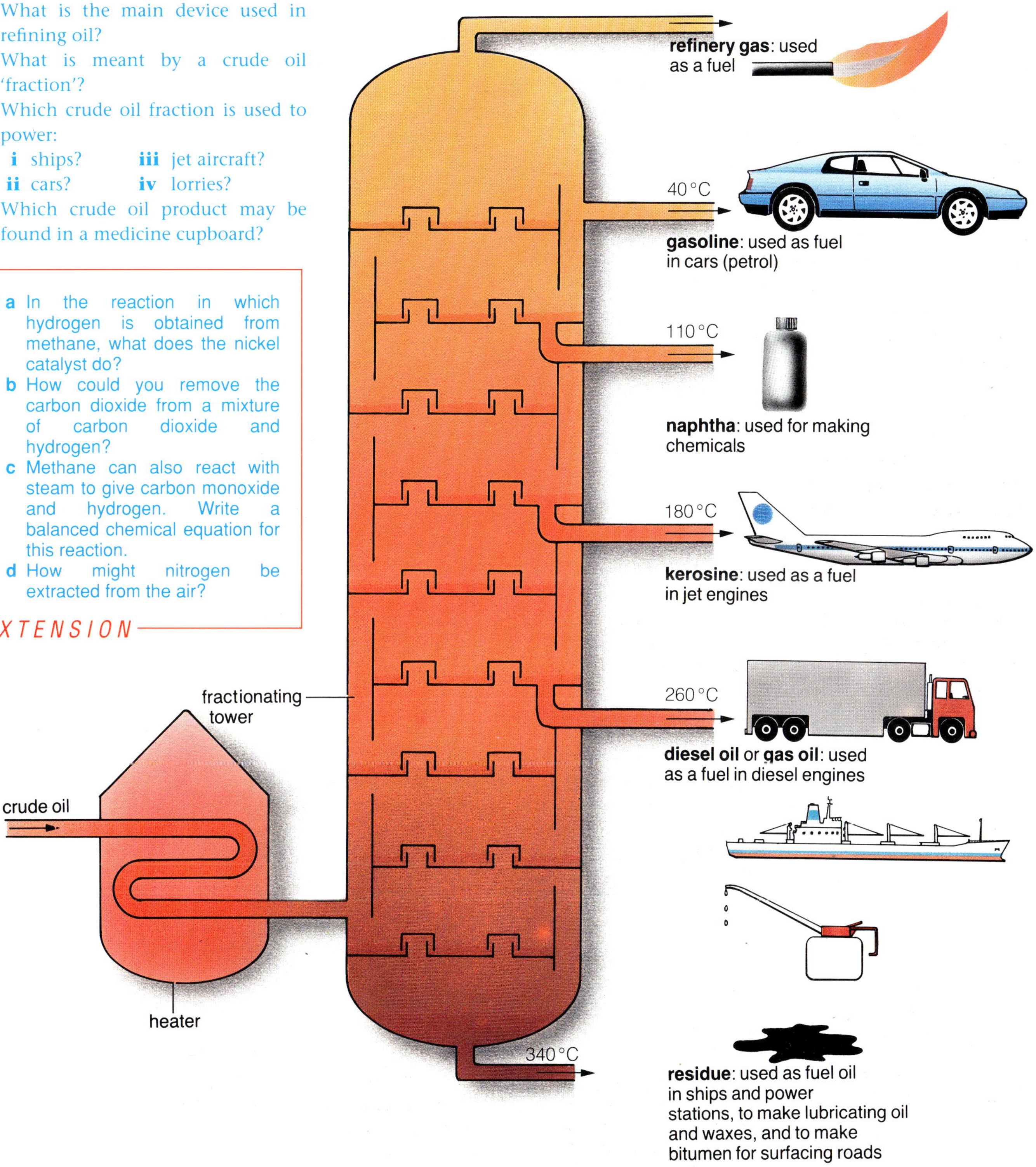

---

refinery gas: used as a fuel

40 °C

gasoline: used as fuel in cars (petrol)

110 °C

naphtha: used for making chemicals

180 °C

kerosine: used as a fuel in jet engines

260 °C

diesel oil or **gas oil**: used as a fuel in diesel engines

fractionating tower

crude oil

heater

340 °C

residue: used as fuel oil in ships and power stations, to make lubricating oil and waxes, and to make bitumen for surfacing roads

Fig. 60.4 Fractional distillation of oil. The hot gases rise up the tower, cooling gradually. As each component cools below its boiling point, it condenses onto the trays. The liquids run off, and are collected.

INVESTIGATION 61.1

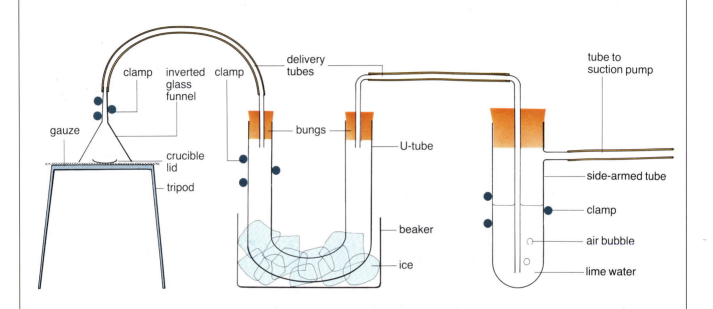

Fig. 61.1

Which elements are present in oil products?

TAKE CARE WHEN DOING THIS EXPERIMENT. WEAR SAFETY GOGGLES.

1 Set up the apparatus shown in the diagram. Make sure that it is all clamped securely! The apparatus is designed to collect the fumes from an oil product burning in the crucible lid. Trace the path followed by the fumes from the glass funnel to the rubber tube connected to the suction pump.

2 Turn on the suction pump to check it is working. You should see bubbles of air being sucked through the lime water.

3 Place a *little* kerosine in the crucible lid. Turn on the suction pump. Light the kerosine. Observe the glass funnel, U-tube and lime water carefully. Record your observations.

4 When the kerosine has finished burning, clean the glass funnel and U-tube and put fresh lime water in the side-armed test tube. Repeat the experiment, using the other oil products your teacher gives you one at a time. Clean the apparatus after each compound.

Natural gas can be obtained from oil, so you can also try the experiment with a Bunsen on a low yellow flame in place of the tripod, gauze and crucible lid.

Questions

1 What did you see forming in the U-tube?

2 Which two elements does this substance consist of?

3 Which of these two elements must have come from the fuel?

4 Where did the other element come from?

5 What happened to the lime water?

6 What compound caused this?

7 Which two elements does this substance consist of?

8 Which of these two elements must have come from the fuel?

9 Was there any other evidence that the fuel contained this element?

10 Use your answers to Questions 3 and 8 to name two elements contained in the fuels.

Hydrocarbon molecules can contain any number of carbon atoms

Most oil products are compounds of hydrogen and carbon. They are called **hydrocarbons**. How can there be so many different hydrocarbons?

Both hydrogen and carbon are non-metals, so they bond covalently. A carbon atom has four outer electrons. Hydrogen has one. One compound which they can form is shown in Figure 61.2. This compound is **methane**.

Fig. 61.2   Methane

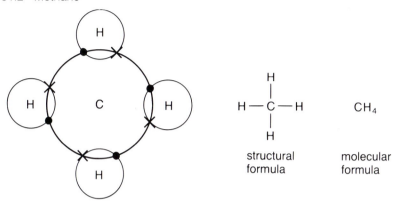

H—C—H CH_4

structural formula   molecular formula

Methane is found in oil. During distillation it comes out at the top of the fractionating tower as part of the refinery gas. It is also the main component of the natural gas you burn in a Bunsen burner.

Another hydrocarbon compound exists which contains two carbon atoms per molecule. The carbon atoms are joined by a single covalent bond. This compound is **ethane**. Like methane, it is a component of the refinery gas.

Fig. 61.3   Ethane

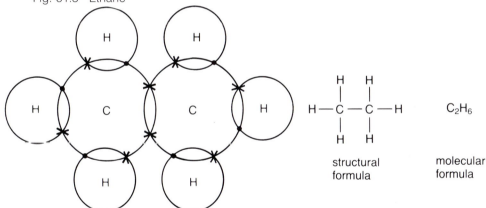

H—C—C—H C_2H_6

structural formula   molecular formula

In fact, any number of carbon atoms per molecule is possible. Three carbon atoms gives **propane**. Four carbon atoms gives **butane**. There is an enormous number of possible compounds, many of which are found in crude oil.

Questions

1 a Draw a structural formula for butane.

b What is the molecular formula of butane?

2 a Write balanced chemical equations for the reactions of methane, ethane, propane and butane with air. (These are all combustion reactions.)

b When these four substances burn, they all contribute to the 'greenhouse effect'.

i What is the greenhouse effect?

ii How do these substances contribute to the greenhouse effect?

iii Do these substances cause any other pollution when they burn?

3 Many people buy butane. In what form is it bought, and what is it used for?

Similar compounds may be grouped together

Methane, ethane, propane and butane all behave in a similar way. They have similar chemical properties. They are quite unreactive compounds. One of the few reactions of which they are capable is combustion in air.

These compounds with similar behaviour and similar composition have been given a group name. They are called **alkanes**. There are other hydrocarbons like them, which also belong to the alkane group.

Although the chemical properties of all the alkanes are very similar, their physical properties do vary. Table 62.1 shows the boiling points of some of the alkanes. You can see that the larger the alkane molecule, the higher its boiling point. This means that large alkane molecules will come out lower down the fractionating column than small ones. Methane, ethane, propane and butane are all part of the refinery gas, because the temperature in the fractionating column never falls below their boiling points. Hexane comes out in the petrol fraction. Hexadecane is in the diesel oil fraction. Alkanes with around 25 carbon atoms per molecule come out in the lubricating oil fraction. The bitumen fraction contains alkanes with over 50 carbon atoms per molecule.

Another physical property which varies amongst the alkanes is viscosity. Alkanes with large molecules are more viscous than alkanes with small molecules.

Table 62.1 The boiling points of some straight chain alkanes

Name	Molecular formula	Boiling point (°C)
methane	CH_4	–162
ethane	C_2H_6	–89
propane	C_3H_8	–42
butane	C_4H_{10}	–0.5
hexane	C_6H_{14}	69
hexadecane	$C_{16}H_{34}$	287

EXTENSION

The alkanes have a general molecular formula

You may have noticed a relationship between the amount of carbon and the amount of hydrogen in each alkane molecule.

Number of H atoms = 2 + (2 x number of C atoms)

So alkanes have the general molecular formula C_nH_{2n+2}. For example: hexane has six carbon atoms, so n = 6.

2n + 2 = (2 x 6) + 2 = 14.

So the molecular formula for hexane is C_6H_{14}.

EXTENSION

The alkanes form a homologous series

A set of hydrocarbons with a group name (like the alkanes) is said to be a homologous series. The following are characteristics of members of a homologous series.

• They have similar chemical properties.
• They have gradually varying physical properties.
• Their formulae fit the general molecular formula.
• Adjacent members, e.g. CH_4 and C_2H_6, or C_5H_{12} and C_6H_{14} differ by one carbon and two hydrogens.

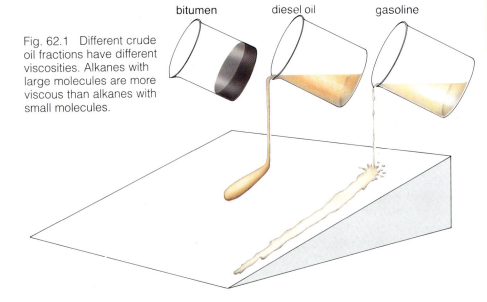

bitumen   diesel oil   gasoline

Fig. 62.1 Different crude oil fractions have different viscosities. Alkanes with large molecules are more viscous than alkanes with small molecules.

As well as enabling them to be separated by fractional distillation, the different boiling points of petrochemicals also affect their ease of burning. Those with small molecules and therefore low boiling points are easy to ignite. This makes them good fuels but also means they are more dangerous. Those with large molecules and therefore high boiling points are harder to ignite. This may make them more difficult to use as fuels but also means they are less likely to cause accidents.

Sometimes the same atoms can bond together in different ways

The fourteen atoms which make up a butane molecule can bond together in two different ways:

Fig. 62.2a   Butane

Fig. 62.2b   2-methyl propane

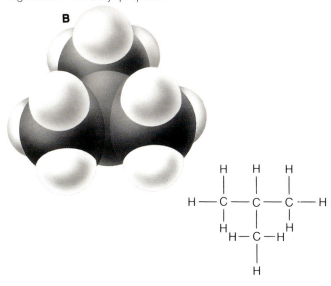

A and B are two different compounds. They are both alkanes. Both have the same molecular formula, C_4H_{10}. Both have very similar chemical properties. But their structural formulae are different, and they have slightly different physical properties. Compound A boils at –0.5°C. Compound B boils at –12°C. Compounds A and B are said to be **isomers**. They are the two isomers of butane.

Note that the difference between these two isomers is that in A the four carbon atoms are in a *straight* chain, whereas in B they are in a *branched* chain. Isomers of hydrocarbons always differ by branching in the chain, not by the chain being bent. Bending does not produce a different isomer. Branching does.

Fig. 62.3   These four structural formulae are all of the same substance – butane.

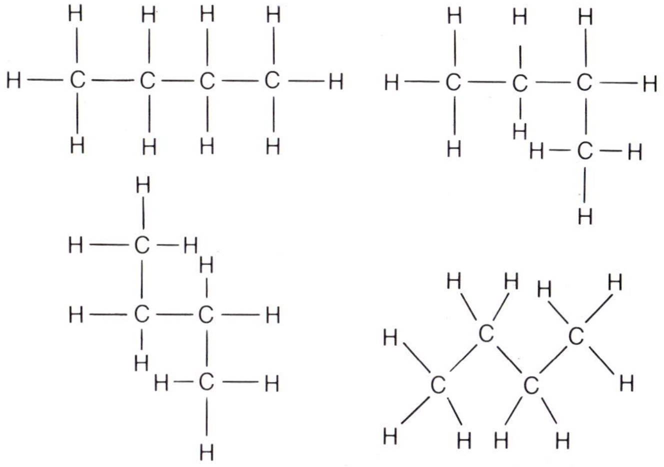

Isomerism has important consequences. If petrol is to burn in the correct way in a car engine, it must be rich in branched isomers, like the isomer of octane, C_8H_{18}, shown in Figure 62.4.

Fig. 62.4   An isomer of octane

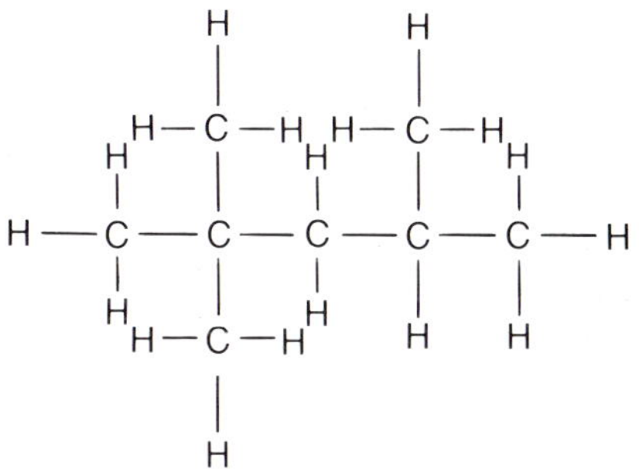

Questions

1 a Estimate the boiling point of the straight chain isomer of pentane, C_5H_{12}.
  b Which crude oil fraction will pentane be part of?

2 a What is the formula of the alkane with:
    i 12 carbon atoms?
    ii 50 carbon atoms?
  b Which fractions will each of these alkanes be part of?

3 Ships run on fuel oil, consisting of alkanes with 30 to 40 carbon atoms per molecule. What problems will arise from this fuel oil's physical properties, and how might these problems be overcome?

4 Which of the following are formulae of alkanes?

  a C_2H_5 d C_9H_{22}
  b C_3H_8 e C_6H_6
  c C_7H_{16} f $C_{10}H_{22}$

5 Draw structural formulae for the three isomers of pentane, C_5H_{12}.

6 If petrol is rich in branched isomers, it does not need to have lead compounds added to it. Explain why such lead-free petrol is a good idea.

63 ALKENES

In addition to alkanes, hydrocarbons of other groups, or homologous series, are also found in crude oil. One such homologous series is the alkenes.

Fig. 63.1 Ethene

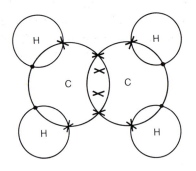

C_2H_4

Alkenes contain double bonds

Alkanes are not the only type of hydrocarbon compound found in oil. Another particularly interesting type has two of the carbon atoms in the molecule joined together by two covalent bonds. This is called a **double bond**. These hydrocarbons are called **alkenes**. The simplest alkene has two carbon atoms, and is called **ethene**.

Compare the structure of ethene with that of ethane on page 149. Notice:
- how the double bond is drawn in the structural formula.
- that since each carbon forms four bonds, each carbon in ethene is joined to two hydrogen atoms, not three as in ethane.
- that 'eth' signifies 'two carbon atoms'. Ethane and ethene both have two carbon atoms per molecule.
- that the names of alkenes end with 'ene'. The names of alkanes end with 'ane'.

Fig. 63.2

propene

C_3H_6

but-1-ene

C_4H_8

but-2-ene

C_4H_8

Note: But-**1**-ene has the double bond in the **first** bond of the chain.

But-**2**-ene has the double bond in the **second** bond of the chain.

INVESTIGATION 63.1

Chemical properties of alkanes and alkenes

You are going to compare the reactions of cyclohexane, and cyclohexene. In these reactions, cyclohexane's behaviour is typical of an alkane and cyclohexene's behaviour is typical of an alkene.

1 WEAR SAFETY GOGGLES. Put two drops of cyclohexane on a crucible lid. Place the crucible lid on a heatproof mat. Light the cyclohexane, and observe carefully how it burns. Repeat with two drops of cyclohexene.

2 Add two drops of cyclohexane to a little bromine water in a test tube. Cork and shake. Observe what happens. Repeat with two drops of cyclohexene.

3 Add two drops of cyclohexane to a little acidified potassium permanganate solution in a test tube. Cork and shake. Observe what happens. Repeat with two drops of cyclohexene.

Questions

1 Suggest a simple method for distinguishing between an alkane and an alkene.
2 What difference did you notice in the flames produced by the alkane and the alkene as they burnt?
3 Which appear to be more reactive – alkanes or alkenes?

152

Polythene molecules are chains of ethene molecules

You have already seen two reactions which are undergone by alkenes but not alkanes. Alkenes decolourise bromine water and acidified potassium permanganate solution. But alkenes have another, more important property.

In 1933, during experiments on ethene conducted under very high pressure, a white, waxy solid was formed. This solid had many useful properties. It could be easily moulded into shape if heated. It did not corrode. It was an excellent electrical insulator. The solid was found to consist of very large molecules, each one made of many thousands of ethene molecules. 'Many' in Greek is 'poly', so the new solid compound was called 'polyethene', or **polythene**.

The diagram of polythene in Figure 63.3 shows only part of a polythene molecule. It shows only seven ethene molecules reacting. But the whole polythene molecule consists of many thousands of ethene molecules in a chain.

The ethene molecules are able to form chains like this because they have double bonds. Ethane molecules cannot form chains. There is no such compound as 'polyethane'. This is because ethane molecules do not have double bonds.

Fig. 63.3 How polythene forms

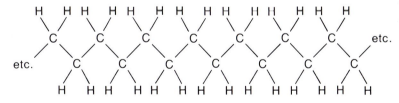

Ethene is a gas, so its molecules are widely spaced.

High pressures force the molecules closer together.

A trace of oxygen in the mixture causes the molecules to string together, forming polythene.

Saturated and unsaturated molecules

Alkanes are said to be **saturated**. This means that each molecule has no carbon-to-carbon double bond and so has the maximum possible number of hydrogen atoms. Alkenes are said to be **unsaturated**. This means that each molecule does have a carbon-to-carbon double bond and so has less than the maximum possible number of hydrogen atoms.

Questions

1 a Write down the structural formulae and the molecular formulae of:
  i pent-1-ene
  ii pent-2-ene.
 b What will happen if each of these two compounds are shaken with bromine water?

2 a What is the relationship between the number of hydrogens and the number of carbons in an alkene molecule?
 b Write the general formula of an alkene, beginning: C_n.

3 Figure 63.2 shows two isomers of butene. Is there a third? Explain your answer.

4 Which of the following compounds could form a polythene-like compound?
  a propene   **d** methane
  b butane   **e** butene
  c propane

5 Is there such a compound as methene? Explain your answer.

64 POLYMERS

Many substances like polythene can be made from alkenes. We call these substances plastics or polymers.

Crude oil provides the raw materials for making plastics

You have seen that polythene is made by forming a long chain of ethene molecules. Many other compounds can be made in a similar way. A compound made like this is called a **polymer**. The compound with small molecules used to make the polymer is called a **monomer**. Polymers like polythene are commonly known as **plastics**. One of the most important types of chemical resource that we get from crude oil is the monomers we use to make polymers. Crude oil is therefore the main source of raw materials for the plastics industry.

Monomers react together in polymerisation reactions

Another alkene which can form polymers is propene. Figure 64.1 shows how it does this. The polymer is called **polypropene**. It is a more rigid, stronger plastic than polythene.

Propene is the monomer from which polypropene is made. The reaction in which monomer molecules join together to form a polymer is called **polymerisation**. Note that in polypropene, as in polythene, the double bonds in the monomer play a key role in the polymerisation reaction.

Another alkene obtained from crude oil is phenyl ethene, commonly called **styrene**. It is a monomer, and can polymerise to form the polymer **polystyrene**.

Polystyrene, polythene and polypropene are all **addition polymers**. Their monomer molecules add together to make polymer molecules. Nothing in the monomer molecules is left out. Addition polymers are made from monomers which have a double bond between two carbon atoms. These monomers are usually alkenes, such as ethene. They may also be compounds based on alkenes, such as chloroethene.

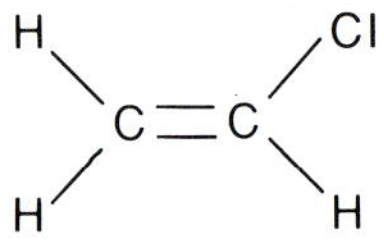

Fig. 64.1   How polypropene forms

Many propene molecules . . .

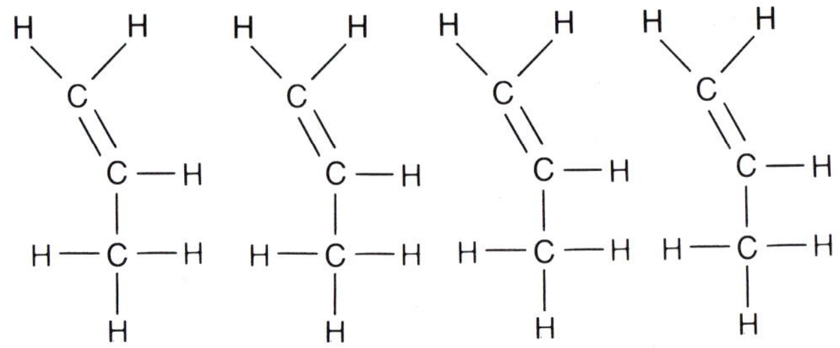

. . . join together to form a long chain.

Fig. 64.2   How polystyrene forms

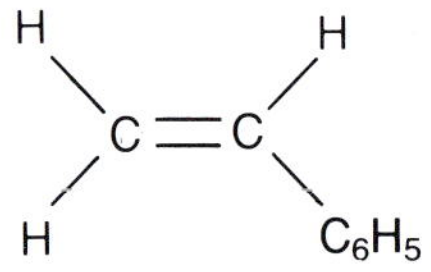

Another alkene obtained from crude oil is phenyl ethene, often called styrene, a monomer . . .

. . . which can polymerise to form polystyrene.

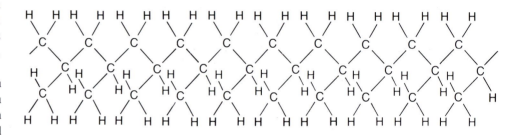

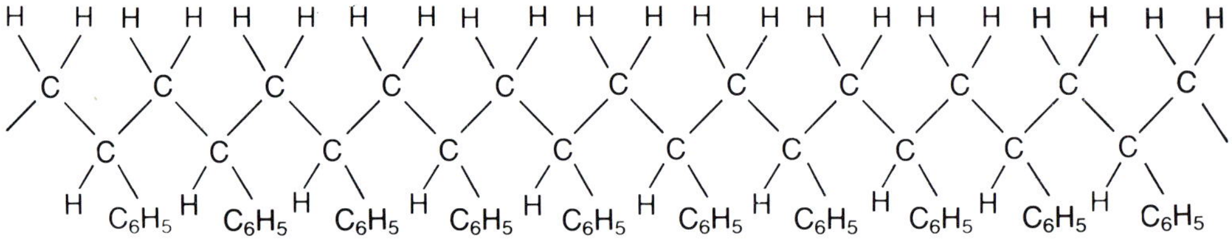

Cracking splits large molecules in crude oil into smaller ones

Crude oil contains a significant amount of the alkenes we use to make addition polymers, but not enough. Most modern oil refineries use some of the less useful crude oil fractions to make more of these useful alkenes. The process used is called **cracking**.

The raw material, sometimes called **feedstock**, might be decane, $C_{10}H_{22}$. Decane is part of the kerosine crude oil fraction. The decane is vapourised and passed through a bed of catalyst powder at around 500°C. The decane molecules split apart, or crack, on the surface of the catalyst grains.

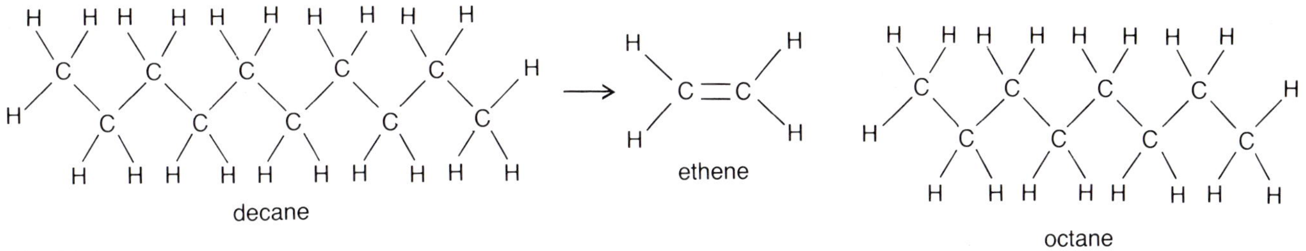

Fig. 64.3a

The products of this process are ethene, which can be used to make polythene, and octane, which is part of the petrol fraction. These products command a higher price than kerosine.

Another useful process which happens on the catalyst surface is called **reforming.** In this process, straight chain molecules like those in octane break up and reform as branched chain molecules.

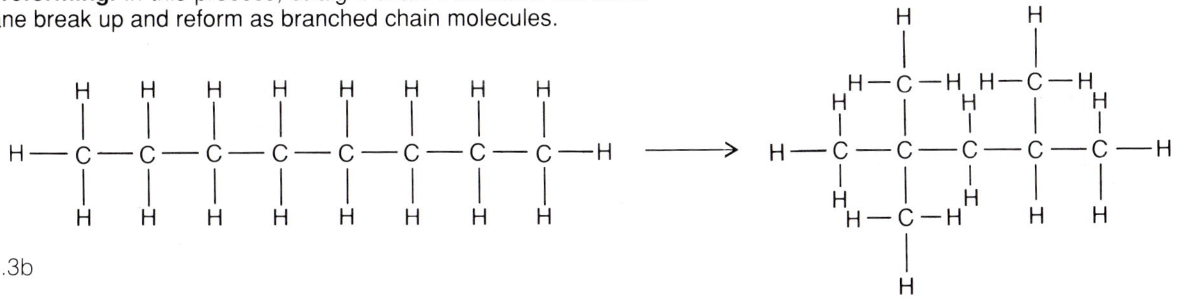

Fig. 64.3b

The branched chain alkanes formed in this process give a much higher performance petrol than straight chain alkanes.

Questions

1 The structural formula for chloroethene is given below.

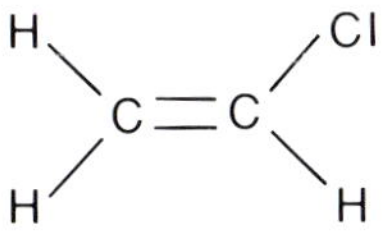

a Draw a diagram to show the addition polymer which can form from this monomer.

b Chloroethene is sometimes called vinyl chloride. What is the polymer called?

2 a Draw a diagram to show the addition polymer formed from the tetrafluoroethene monomer below.

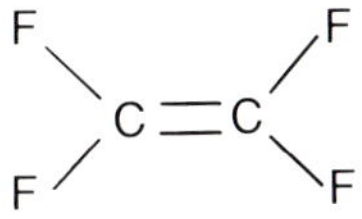

b Why do you think this polymer is known as PTFE?

3 a Is propene a saturated or an unsaturated compound? Explain your decision.

b Is polypropene a saturated or an unsaturated compound? Explain your decision.

4 Dodecane, $C_{12}H_{26}$, can be cracked. The dodecane molecules may split in a number of ways. What is the other product formed, if each dodecane molecule splits to give:

a an ethene molecule?

b an octane, C_8H_{18}, molecule?

c a propene molecule?

d a hexane, C_6H_{14}, molecule?

155

65 CONDENSATION POLYMERS

Condensation polymers contain a different sort of linkage from addition polymers.

EXTENSION

Nylon is made from oil products

Most of the substances we know as 'plastics' are addition polymers, made from alkenes. But there is also another important type of polymer – the **condensation polymer**. The first totally synthetic condensation polymer was made in 1938. The monomers again came from crude oil products. But this time, there were two different monomers:

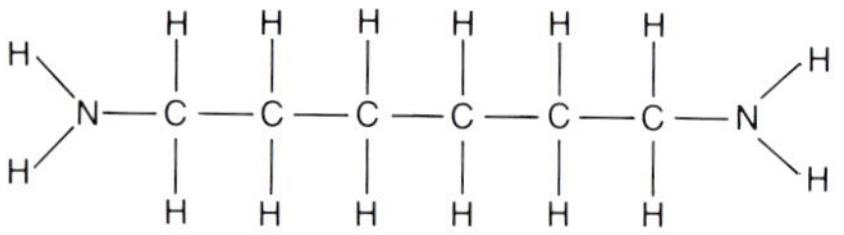

1,6 – diaminohexane

and

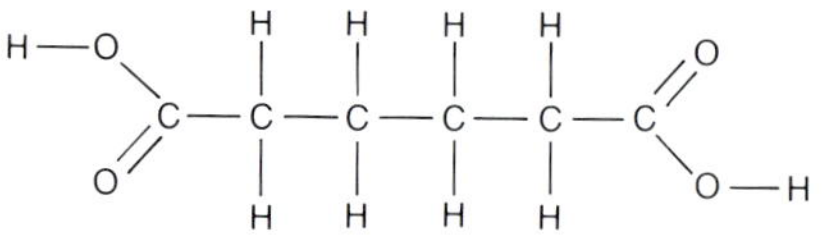

hexanedioic acid

These two molecules can join together, because they have 'reactive ends'.

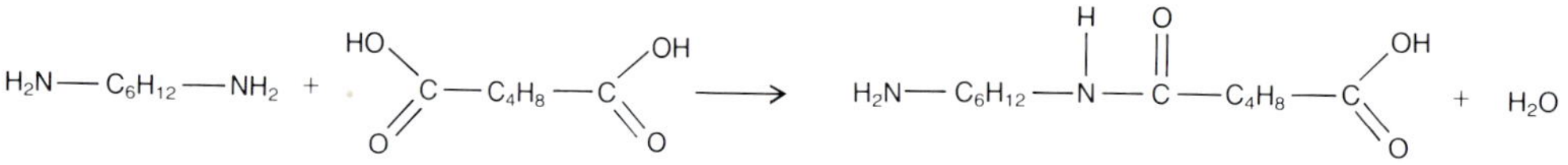

The product of this reaction still has two reactive ends. The product can react like this:

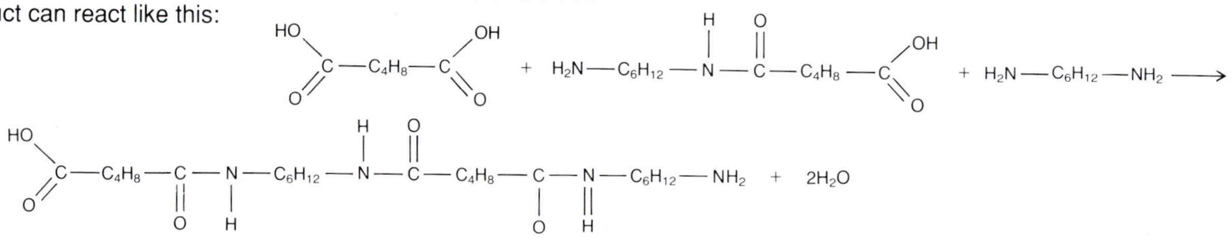

The product *still* has two reactive ends. So the process can go on and on, producing a very long chain molecule. This long chain molecule is a polymer, made from two monomers. It is called a **condensation** polymer, because each time a monomer bonds onto the growing chain, a water molecule is produced. The water may **condense** as droplets in the reaction vessel.

This polymer is a type of polyamide. This particular one was named **nylon**. It is used to make many things, including clothing and fishing lines.

Polyester is also a condensation polymer

Another important condensation polymer is polyester. Again, there are two different monomers, both of which can be made from crude oil products.

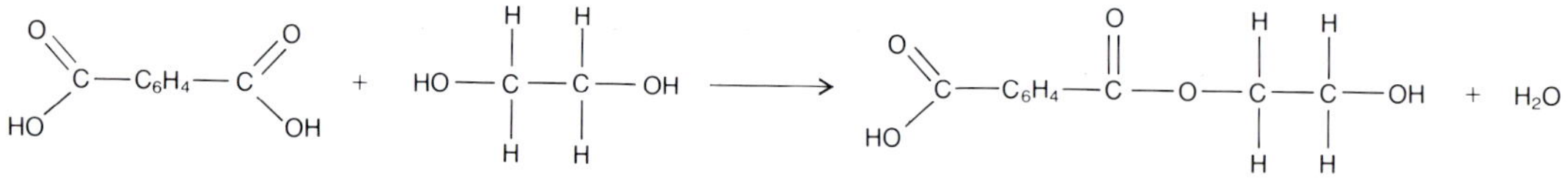

As you can see, this is very similar to the first step in nylon formation. The monomers have reactive ends. A water molecule is produced as the monomers join together. The product still has two reactive ends, so the chain can continue to grow, becoming:

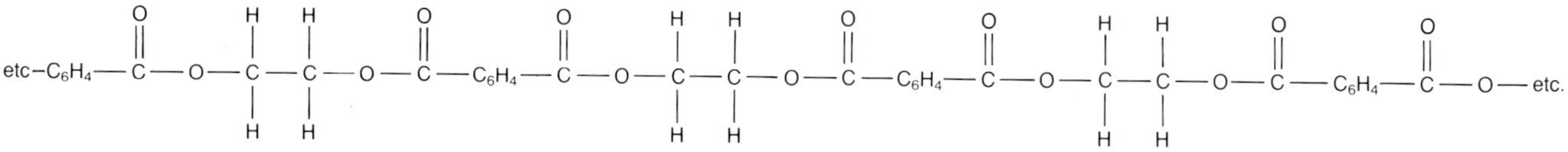

This particular polyester is known as **Terylene.** It was first made in 1941. Like nylon, Terylene is much used to make artificial fibres for clothing. But it would be wrong to assume that all artificial fibres are made from condensation polymers. The acrylic fibres which are now very popular as a wool substitute are made from an addition polymer.

Properties of polymers

You will be given samples of several different polymers, plus a few of the substances which have been replaced by polymers. You should test these substances straight after testing the polymer which has replaced them. For example, test paper straight after testing polythene sheet. Test wool after testing acrylic fibres. This makes it easier to compare them.

1 Place a piece of each substance in a little dilute hydrochloric acid in a labelled test tube. Leave them while you do the rest of the experiment.

Later, look to see if there is any evidence that any of the substances have reacted, or have been corroded.

2 In a fume cupboard, heat a small piece of each substance directly in a Bunsen flame. AVOID BREATHING VAPOURS OF BURNING POLYMERS. Ask yourself particularly:

- does it burn easily?
- if it burns, is the flame clean or sooty?
- does it keep on burning if removed from the flame?
- does it melt or drip?

3 If your sample is of a suitable size and shape, test its elasticity. To do this, make a loop from the wool and from the acrylic samples. Cut a strip right around the paper bag and the polythene bag. (What must you do to make sure that the comparison is fair?) Then hang the loop of your sample material from a rctort stand and clamp. Hang your smallest mass from the loop. Measure the length of the loop. Gradually add mass to the loop, recording the length for each mass.

Compare acrylic with wool, and polythene with paper.

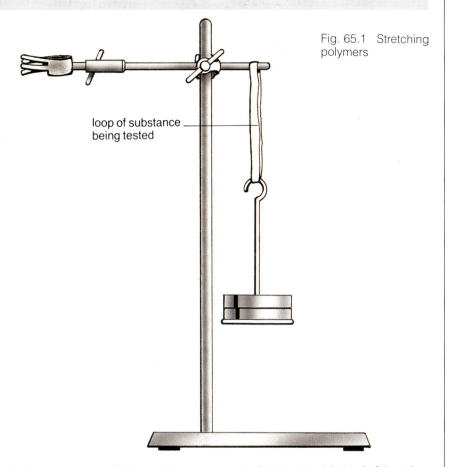

loop of substance being tested

Fig. 65.1 Stretching polymers

Which pair is most similar? Why should this be? Is either polymer very superior to the substance it replaces?

4 Make some strong brine solution and boil it. TAKE CARE. WEAR SAFETY GOGGLES. Adjust your Bunsen so that the brine 'simmers'. Add a small piece of each substance to the boiling brine. Keep a similar sized, untreated, piece for comparison. Simmer for five minutes. Examine the substances.

Has the heat affected them? Have they softened, shrunk or shrivelled? What is the significance of this?

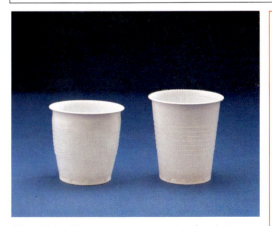

Fig. 65.2 These cups are made of polystyrene sheet. The one on the left has boiling water in it, which has caused the polystyrene to soften and change shape.

Questions

1 List the differences between addition polymerisation and condensation polymerisation.
2 The type of nylon on the opposite page is called 66 (six six) nylon because the two monomers both have six carbon atoms. Another nylon is called 610 nylon. What do you think 610 means?
3 6 nylon has only one type of monomer, yet monomer units link in the same way as in other nylon types. Suggest how this might be possible, and draw a possible structure for the monomer.
4 Acrylic fibres are made from an addition polymer, the monomer of which is propene nitrile:

How does this monomer polymerise? What is the structure of the polymer molecule?

EXTENSION

66 PROPERTIES AND USES OF POLYMERS

Polymers have a wide variety of uses.

Polymers have many useful properties

Polymers have properties which make them very suitable for all sorts of uses.

- They do not corrode.
- They have good strength and elasticity properties.
- They are electrical and thermal insulators.
- They can be produced in different colours.
- They have a low density.
- They are easily moulded – this is what 'plastic' actually means.

Different polymers show these properties to different degrees.

Some polymers can be remoulded as often as desired after first softening them by heating. These are called **thermoplastic** polymers. This is because the forces between neighbouring polymer chains are weak, so the polymer can easily soften or melt when heated. Polythene and polystyrene are examples of thermoplastic polymers. Other polymers soften and can be moulded the first time they are heated, but cannot be resoftened and remoulded. These are called **thermosetting** polymers. This is because strong covalent bonds form between neighbouring polymer chains the first time the polymer is heated. If the polymer is heated on future occasions, these bonds hold the chains in their existing positions. Phenolic resins, which are polymers used to make electric plugs and sockets, are examples of thermosetting polymers.

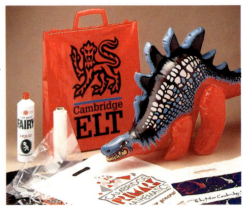

Fig. 66.1 **Polythene** is very cheap and is easily moulded into strong, flexible containers.

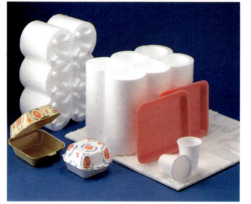

Fig. 66.2 **Polystyrene**, like polythene, is cheap and easily moulded. Expanded polystyrene is made by blowing gas into the polymer while it is setting. This makes a very light, insulating packing material.

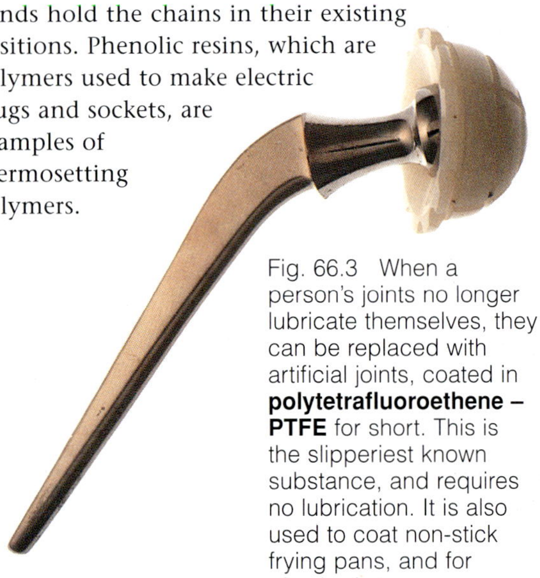

Fig. 66.3 When a person's joints no longer lubricate themselves, they can be replaced with artificial joints, coated in **polytetrafluoroethene – PTFE** for short. This is the slipperiest known substance, and requires no lubrication. It is also used to coat non-stick frying pans, and for plumber's tape.

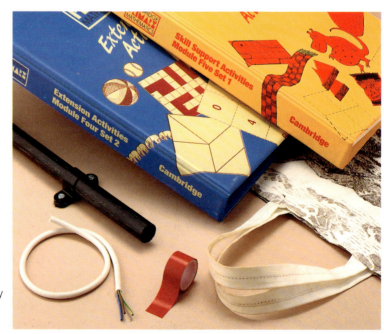

Fig. 66.4 **Poly vinyl chloride, PVC**, is flexible and shiny. It is used for making clothing, and waterproof coatings for bags and folders. It is a good electrical insulator, so it is useful for making insulating tape or coverings round electrical wiring. For making windows and piping, uPVC, which is more rigid, is used. Unfortunately, PVC can cause a pollution problem when it is disposed of, as it produces half its own weight in hydrochloric acid when burnt.

Fig. 66.5 **Nylon** is easily moulded into fibres and has many attractive properties, including strength, elasticity and long life. As well as its best known use, in clothing, it is also used to make gear wheels, ropes and fishing lines.

Fig. 66.6 Strong mouldable, shatterproof and transparent **Perspex** is ideal for this cockpit covering. The chemical name for Perspex is **polymethylmethacrylate.** It has replaced glass in many situations. You probably have Perspex safety screens in your school lab.

Fig. 66.7 Plastic pollution at Kimmeridge, Dorset. Plastic articles are washed up onto beaches after being discarded from ships. They do not decay, and will stay on the beach for a very long time unless someone clears them away.

Fig. 66.8 Polyurethane foams were, until recently, often used to fill furniture. Their flammability and toxic combustion products contributed to many deaths in house fires.

Polymers do have some disadvantages

Polymers are widely used because they either have better properties than the natural alternative, or because they are cheaper. You are probably aware that a nylon shirt is much cheaper than a silk one, and that PVC-covered car seats are much cheaper than leather-covered ones. But people still buy silk clothing and leather-covered furniture, because they prefer the natural look and feel. They are prepared to pay the extra price. The 'artificial' appearance of polymers can be a disadvantage in some situations.

A second disadvantage of polymers is that many of them do not rot down. Natural substances can be broken down by bacteria and fungi. They are said to be **biodegradable**. But most polymers are **non-biodegradable**. A polythene bag thrown under a hedge will still be there in many years' time, long after a paper one would have rotted away. Biodegradable plastics are now being developed to try to prevent this pollution problem in the future.

A third disadvantage is the fire risk associated with some polymers. Some of them catch fire easily. Others give off dangerous fumes when they burn. Polyurethane foams were, until very recently, used to fill furniture. If they catch fire, they give off poisonous gases. A number of deaths in house fires have been caused by these gases.

Questions

1 Explain why the following are often made of polymers:
   a telephones
   b toys
   c carrier bags
   d ball point pens.
2 Explain why the following are made of thermosetting polymers:
   a saucepan handles
   b electric plugs.

3 Polymers may eventually become scarce. Discuss the possible reasons for this. What other resources will then be used in greater quantity?
4 Why does the recycling of polymers make good sense? Why does very little polymer recycling take place at the present time?

EXTENSION

Limestone is obtained by quarrying

In many parts of the UK there are large deposits of limestone very close to the surface. The limestone is quarried – this means we dig straight into it from the surface leaving a large empty pit. As the pit gets bigger, roads are built into it so trucks can drive in to carry away the limestone. The pit is extended outwards by drilling holes into its sides, and explosives are then put into the holes causing more broken limestone to fall into the pit.

Limestone is basic

Limestone is impure calcium carbonate, $CaCO_3$, and so it will neutralise acids. It is added to blast furnaces to neutralise acid impurities in the iron being made. In Scandinavia, limestone is crushed and the powder is sprayed into lakes that have become polluted by acid rain. The polluted lakes have a pH of below 5, and the limestone treatment raises the pH above 5 so fish and other organisms are able to live in the lake again.

If limestone is heated strongly it decomposes to produce lime (calcium oxide, CaO) and carbon dioxide:

$$CaCO_3 \longrightarrow CaO + CO_2$$

This is done in a large rotating kiln called a lime kiln.

Lime reacts with water to produce slaked lime (calcium hydroxide, $Ca(OH)_2$).

$$CaO + H_2O \longrightarrow Ca(OH)_2$$

Slaked lime is a metal hydroxide, so it is also basic. It is reasonably soluble in water, so it is an alkali. Slaked lime can be added to soils if they are too acidic. This also adds calcium, which plants need. Slaked lime is added to water at water treatment plants if the supply is too acidic. As well as neutralising the acid in the water it adds calcium to the water, which is good for the teeth and skeletons of those who drink it. Unfortunately it also makes the water 'hard', so that when soap is used it doesn't lather as well.

Fig. 67.1 A limestone quarry

Fig. 67.2 Liquid concrete being poured over steel reinforcements.

Limestone is used to make cement

If the limestone is mixed with clay before heating in the rotary kiln, the mixture of lime and roasted clay that forms is called **cement.** Cement is mixed with sand and water to make **mortar** which is used to 'glue' the bricks together in housebuilding. If cement is mixed with sand, small stones and water, the mixture reacts together and hardens to form concrete. In terms of the amounts used, concrete is the most important building material in the developed world. Concrete can be strengthened by the addition of steel reinforcing bars before it sets – and once again, limestone is involved in the manufacture of steel.

Glass is made from limestone and other common minerals

The commonest form of glass, **soda-lime glass**, is made by heating together a mixture of sand (silicon dioxide, SiO_2), limestone (calcium carbonate, $CaCO_3$) and soda (sodium carbonate, Na_2CO_3). Vast deposits of soda are found around East African lakes like Lake Magadi and Lake Naivasha, but most of the soda used for glass-making is made by the **Solvay process**. This uses salt (sodium chloride), ammonia and carbon dioxide as reactants. At the high temperatures of the glass-making process, the limestone and soda break down to give calcium oxide and sodium oxide.

$$CaCO_3 \longrightarrow CaO + CO_2$$

$$Na_2CO_3 \longrightarrow Na_2O + CO_2$$

These two substances then react with the silicon dioxide to form glass in its molten state at temperatures well over 1000 °C.

The molten glass can now be blown into shapes by glassblowers. If the molten glass is floated on top of a bed of molten tin it forms large, perfectly flat sheets called float glass. Float glass is used to make plate glass windows, because the flat surface ensures that things you see through the glass are not distorted.

Fig. 67.3   These gas jets are used to heat the raw materials in the glass furnace.

Fig. 67.4   The finished float glass is carried on rollers to be washed and cut.

Questions

1 a List as many uses of limestone as you can.

b How many of these uses depend on limestone being basic?

c Why is limestone so important to the building industry?

2 a Would you like a limestone quarry near your home?

b Why?

c When all the limestone has been dug out, the empty pit might be used for dumping rubbish (a 'municipal tip') or it might be flooded and used for water sports like water-skiing, sailing, and fishing. Which would you prefer?

d Why?

e Why are bushes and long grasses encouraged to grow around the edges of many flooded quarries?

3 a What substances is glass made from?

b Where does each of these substances come from?

c Apart from these substances, what are the main costs of making glass?

4 Chemical plants which do different things are often built close together. This is particularly helpful when the product (or sometimes even the waste product) of one process is needed by the other. What is the advantage in the following processes being close together?

a Solvay process and glass-making

b limestone quarry and blast furnace

c Solvay process and Haber process

d Haber process and fertiliser manufacture

e lime kiln and Solvay process.

Metals can be extracted from ores

One of the reasons for the success of the human species on Earth is that humans can use tools. They acquired this ability many hundreds of thousands of years ago. Tools made of chipped stones have been found with the very earliest fossil human bones. Early humans also used bones and antlers as tools.

Humans would have known about metals a very long time ago, because some metals are found pure, or **native**. One such metal is **gold**. Gold is too soft for making tools, but it was much used for making jewellery.

Small amounts of **copper**, which was a much more useful metal, were probably also found native. But these early humans did not realise that there was much more copper close at hand. The copper was combined with other elements in the form of copper compounds such as copper carbonate, $CuCO_3$. Copper carbonate is found in the mineral **malachite**. Minerals or rocks such as malachite, which contain metal compounds, are called **ores**.

We do not know how people first discovered how to get copper out of copper ores. But it must have involved fire and luck. Soon afterwards, humans discovered how to get tin from tin ores, also using fire. Mixing and melting copper and tin made the alloy, **bronze**.

Bronze was really useful. It was not brittle, and it could be sharpened. So it provided better tools, and better weapons.

The discovery of **iron** made an even bigger difference to people's lives. Better, hotter furnaces were needed to extract iron from its ores. Iron is harder, stronger, and keeps its edge better than bronze. Iron made much better tools and weapons.

Archaeologists studying the history of the human species often use the terms 'The Stone Age', 'The Bronze Age' and 'The Iron Age' to describe these different stages in the use of metals. These stages happened at very different times in different parts of the world. In Britain, the Bronze Age began around 1100 BC. The Iron Age began around 1000 BC, when the Celts began to settle in Britain, bringing their knowledge of iron with them. In Egypt, bronze was already in use before 2500 BC and in some parts of the world, such as New Guinea, people had still not discovered the use of metals even in the nineteenth century.

Today, we are still in the Iron Age! Over 90% of the metal we use is iron.

Fig. 68.1 The earliest human tools were made from natural materials such as stone and bones. These flint tools are about 4000 years old, and were found in England. From the top, there are three leaf-point arrow heads, a polished hand-axe, a spear head and two arrow heads.

Fig. 68.2 Gold has always been prized for making jewellery and other decorative objects, because it does not corrode. It can be found as a pure metal, so has been used by humans for a very long time. This gold stag is from a tomb in the Crimea, and dates from 5000 BC.

Fig. 68.3 The discovery of how to make bronze made a very big difference to people's lives because it could be used to make sharp and quite strong tools and weapons. It is also a very decorative alloy. This 3500-year-old bronze stand, from Cyprus, shows a man carrying an ingot of copper or bronze. Ingots like these have been found in the wreckage of ancient ships in the Mediterranean.

Some metals are more reactive than others

Metals are amongst our most precious resources. But many metal ores are in short supply. In particular, the ores which most readily give up their metals, like copper, tin and lead ores, are running out.

Why is it easier to get some metals from their ores than others? It depends on the **reactivity** of the metal.

A reactive substance is one which readily takes part in chemical reactions. Gold, copper, tin, iron and lead are all elements. When they react, they form compounds. A reactive metal forms compounds easily. An unreactive metal forms compounds less readily. An understanding of the reactivity of different metals helps us to understand how to extract them from their ores, and also explains why they have different uses.

You can investigate the reactivity of metals in a number of different ways. One method is to look at the reaction between a metal and an acid, to form a salt. For example, if pieces of magnesium, iron and copper are put into separate tubes of dilute hydrochloric acid, they may react. Bubbles of hydrogen gas will be produced. Figure 68.4 shows this. From the rate of bubbling, it is clear that magnesium is reacting faster than iron, and that copper is not reacting at all. This is evidence that magnesium is more reactive than iron, and that iron is more reactive than copper.

Fig. 68.4   Magnesium (left) reacts quickly in hydrochloric acid, iron (centre) reacts slowly, and copper (right) does not react at all.

Fig. 68.5   Crystals of malachite, which contains copper carbonate. Malachite is a valuable ore of copper.

Fig. 68.6   Wheal Betsy, in Devon, stopped producing tin many years ago. This is the engine house above the mine. As tin becomes scarcer and more valuable, some of the old mines in the south west of Britain may be reopened.

INVESTIGATION 68.1

The reaction of metals with acids

WEAR SAFETY GOGGLES

1 Put 2 cm depth of dilute hydrochloric acid into each of five clean test tubes.

2 At the same time (or as close as possible), drop a piece of magnesium, iron, copper, lead and zinc into different tubes. The pieces should be about the same size.

3 Observe the rate of fizzing. Decide on the order of reactivity.

Questions

1 What is an ore?

2 Explain why the Bronze Age came before the Iron Age.

3 Give three uses for which metals are essential.

4 a Which word describes elements that do not readily form compounds?

b Which word describes elements that readily form compounds?

5 a Write balanced equations for the reactions of zinc, iron and magnesium with hydrochloric acid.

b Are the salts formed by these reactions ionic or covalent compounds?

c In the reaction between magnesium and hydrochloric acid, identify the substance which is oxidised, and the substance which is reduced.

6 Magnesium is a very difficult metal to extract from its ore. Do you see any relationship between reactivity and ease of extraction? Think particularly about iron, magnesium and copper.

Some metals react with water

If you have looked at the reactivity of metals with acid, you should now know this reactivity series:

> magnesium
> zinc
> iron
> copper or lead.

Other common metals also react with acids, but too violently for these reactions to be carried out safely. To compare the reactivity of these metals with the five metals in the reactivity series above, the reaction with water to form metal hydroxides can be studied. The extra metals which you can look at

Fig. 69.1   From left to right: lithium, calcium and magnesium reacting in water.

in this way are lithium and calcium.

If clean samples of magnesium, zinc, iron, copper and lead are dropped into water, only the magnesium will be seen to be reacting a minute later. It reacts

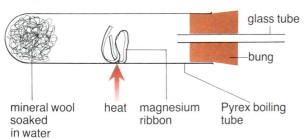

glass tube

bung

mineral wool soaked in water   heat   magnesium ribbon   Pyrex boiling tube

Fig. 69.2   Apparatus for reacting magnesium with steam. (Figure 85.1 shows a photograph of this experiment.)

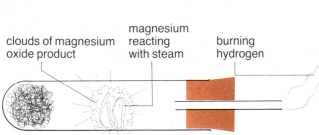

clouds of magnesium oxide product   magnesium reacting with steam   burning hydrogen

very slowly. The magnesium ribbon will eventually float, because hydrogen bubbles are formed which stick to it. The other product is magnesium hydroxide.

Now look at the photographs of lithium and calcium reacting in water (Figure 69.1). Calcium reacts rapidly, giving off a steady stream of hydrogen bubbles. Lithium reacts even more rapidly: hydrogen is given off fast enough to burn with a steady flame once lit.

So we can now add these two metals to the reactivity series:

> lithium
> calcium
> magnesium
> zinc
> iron
> copper or lead.

Metals which won't react with cold water may react with steam

Just as acid was no help in deciding whether copper or lead is the more reactive, cold water is no help for zinc, iron, copper or lead. But reactions speed up if you heat them. How do these metals react with very hot water – steam?

We find that copper and lead do not react even with steam, but zinc and iron do react with steam, and so does magnesium. The products formed when a metal reacts with steam are hydrogen and a metal oxide.

Figure 69.2 shows the apparatus used to react a metal with steam. In this diagram, magnesium is being reacted with steam. The magnesium is heated – not the mineral wool. But the heat still produces some steam from the soaked wool. When the magnesium is very hot, it starts to react. It does so

with a glaring yellow/white light. Hydrogen gas is produced, and burns fiercely if a flame is put to it.

Zinc undergoes the same reaction, but less readily. The zinc needs to be hotter before it will react. It also needs a continuous flow of steam over it. Even then, it does not flare up as magnesium does. The hydrogen is not produced as quickly.

Iron reacts with steam only when the metal is white-hot. Even then, there is only a very slow reaction.

These reactions with steam confirm the order of reactivity:

> magnesium
> zinc
> iron.

Some metals burn in air

Many metals burn when heated directly in air. There is only one product of this reaction – the oxide of the metal used. These reactions are too quick to be of much use in judging metal reactivities, but the results do fit roughly into the reactivity series worked out so far. Some of these reactions are shown in Figure 69.3.

The less reactive metals do not burn when heated in air. A layer of black copper (II) oxide will, however, form slowly on the surface of copper when it is heated.

This confirms the reactivity series:

> lithium
> calcium
> magnesium
> aluminium
> zinc
> iron
> copper or lead.

Questions

1 a Write word equations for the reactions of the following metals with water:
  i lithium
  ii calcium
  iii magnesium.
  b Write balanced equations for reactions **i** and **ii**.
  c How do the reactions of lithium and calcium with water give evidence that lithium is more reactive than calcium?
2 a Write word equations for the reactions between steam and:
  i magnesium
  ii zinc.
  b Write balanced equations for these two reactions.
  c Why should you not try to react lithium or calcium with steam?
3 a Give the names and formulae of the compounds formed when magnesium and lithium burn in air.
  b What is the formula of the compound formed when aluminium burns in air?
  c Why do some metals burn as powders when they will not burn as large pieces?

Fig. 69.3a Lithium burns vigorously.

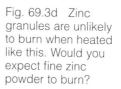

Fig. 69.3b Calcium granules about to begin their sudden combustion.

Fig. 69.3c Magnesium ribbon burning.

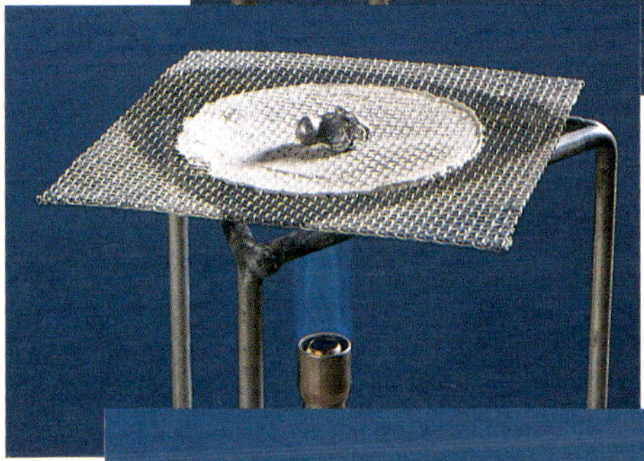

Fig. 69.3d Zinc granules are unlikely to burn when heated like this. Would you expect fine zinc powder to burn?

Fig. 69.3f Fine iron filings burn very vigorously when sprinkled into a blue Bunsen flame.

Fig. 69.3e Iron granules will eventually glow red, but will not burn when heated like this.

Fig. 69.3g Copper blackens when heated, but does not burn.

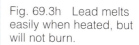

Fig. 69.3h Lead melts easily when heated, but will not burn.

Reactive metals can displace less reactive metals from their compounds

When magnesium is heated in steam, hydrogen is formed.

magnesium + steam → magnesium oxide + hydrogen
$$Mg(s) + H_2O(g) \rightarrow MgO(s) + H_2(g)$$

We say that the magnesium has **displaced** the hydrogen. The magnesium has taken the place of the hydrogen in the oxide compound.

You will remember that a reactive element forms compounds more readily than an unreactive one. In this reaction, magnesium is becoming part of a compound, magnesium oxide. The displaced hydrogen is leaving its compound, water. This shows that magnesium is more reactive than hydrogen.

Displacement reactions can also occur between metals. For example:

copper (II) oxide + magnesium → magnesium oxide + copper
$$CuO(s) + Mg(s) \rightarrow MgO(s) + Cu(s)$$

This is a very vigorous reaction! The copper is displaced by the magnesium. This confirms the results of the reactions on the previous two pages – magnesium is more reactive than copper. Just to check, we can heat magnesium oxide with copper. Nothing happens. Magnesium cannot be displaced by copper.

Fig. 70.1 Magnesium reacting with copper (II) oxide.

Fig. 70.2 The Thermit reaction in use. Molten iron has formed and is flowing down to fill the gap between two rails at the bottom.

Displacement reactions tell you which one of two metals is the more reactive

Displacement reactions give much more positive evidence about reactivity than comparing the reactions of metals with water or acid. If metal A displaces metal B, then metal A is more reactive than metal B. If metal B displaces metal A, then metal B is the more reactive of the two.

We can now, at last, sort out the relative reactivities of lead and copper! If lead oxide is heated with copper, nothing happens. If copper (II) oxide is heated with lead, lead oxide is produced. So lead displaces copper from the copper compound. Lead is more reactive than copper.

One famous displacement reaction demonstrates that aluminium is more reactive than iron. It is called the 'Thermit' reaction. Aluminium powder is mixed with iron (III) oxide powder. The mixture is heated to start the reaction. Once started, the reaction is very rapid. A large amount of heat is given off. The products are aluminium oxide and some very hot iron. This reaction has been used to produce molten iron, to weld lengths of railway line together.

Solutions of salts can be used to perform displacement reactions

Displacement reactions like the Thermit reaction can be very exciting, but they are also dangerous. The Thermit reaction produces molten iron, which can do a lot of damage. A safer way of performing displacement reactions in the laboratory is by using solutions of metal salts, instead of the metal oxide. A piece of a pure metal is put into a solution of the metal salt of a different metal. If the pure metal is more reactive than the metal in the salt, then the metal in the salt is displaced. The metal in the salt appears in the liquid as a pure metal, often as whisker-like growths on the surface of the pure metal you added originally.

For example, copper is less reactive than magnesium. If a piece of magnesium is put into a beaker containing some copper (II) sulphate solution, the magnesium displaces the copper from its compound. Some of the displaced copper sticks to the magnesium, while some falls to the bottom of the beaker. The magnesium ribbon slowly disappears, as the magnesium takes the place of copper in the solution – magnesium sulphate is formed.

magnesium + copper (II) sulphate → magnesium sulphate + copper

$$Mg(s) + CuSO_4(aq) → MgSO_4(aq) + Cu(s)$$

Of course, if you try putting a piece of copper into some magnesium sulphate solution, nothing will happen. Copper cannot displace magnesium.

Fig. 70.3b When the magnesium is taken out after a minute, copper is clearly visible.

Fig. 70.3a Magnesium is added to copper (II) sulphate solution.

Displacement reactions and redox

In all displacement reactions, oxidation and reduction takes place. Metallic elements lose electrons, and form compounds. They are oxidised. Other metals, in compounds, gain electrons, and become pure elements. They are reduced.

For example:

copper (II) oxide + magnesium ⟶ magnesium oxide + copper
$$Cu^{2+}O^{2-}(s) + Mg(s) ⟶ Mg^{2+}O^{2-}(s) + Cu(s)$$

Half equations:

$Cu^{2+} + 2e^- ⟶ Cu$ Copper is reduced
$Mg - 2e^- ⟶ Mg^{2+}$ Magnesium is oxidised

Another example:

zinc sulphate + magnesium ⟶ magnesium sulphate + zinc
$$Zn^{2+}SO_4^{2-}(aq) + Mg(s) ⟶ Mg^{2+}SO_4^{2-}(aq) + Zn(s)$$

Half equations:

$Zn^{2+} + 2e^- ⟶ Zn$ Zinc is reduced
$Mg - 2e^- ⟶ Mg^{2+}$ Magnesium is oxidised

— E X T E N S I O N —

Questions

1 When nickel is heated with copper (II) oxide, nickel oxide and copper are formed.
  a Which of the two metals is more reactive?
  b What other experiments could you do to check on this?
2 Which of the following mixtures would you expect to react when heated? Give products for those which react.
  a zinc and copper (II) oxide
  b zinc and lithium oxide
  c copper and iron (III) oxide
  d magnesium and calcium oxide
  e lithium and calcium oxide
  f iron and lead (II) oxide
3 **a** You are given samples of manganese and iron, and samples of their soluble salts manganese (II) chloride and iron (II) chloride. How could you find out which metal is more reactive?
  b Manganese is more reactive than iron. What would you see happening if you performed the experiment you suggest in your answer to part a?
4 Why would you not try to prove that lithium is more reactive than magnesium by putting lithium into magnesium sulphate solution?

5 **a** Write a word equation for the Thermit reaction.
  b Write a balanced equation for this reaction.
  c Write half equations for the metals involved, and identify the metal which is oxidised and the metal which is reduced.

— E X T E N S I O N —

71 REACTIVITY AND METAL EXTRACTION

Metals can be extracted from their ores by heating the ore with a more reactive element.

Unreactive metals are the easiest to extract from their ores

The displacement reactions described in Topic 70 confirm this reactivity series:

most reactive
lithium
calcium
magnesium
aluminium
zinc
iron
lead
copper
silver
least reactive

You may already have realised that the first metals to have been used by humans were unreactive ones, like copper and lead. These are the easiest metals to extract from their ores. Those metals which we have only begun to use in the last 200 years or so are more reactive metals, like aluminium and sodium. These are more difficult to extract from their ores.

The reason for this is that reactive metals form compounds more readily than unreactive metals. So unreactive metals leave their compounds more readily. Ores are compounds of metals. Unreactive metals leave their ores more easily than reactive ones.

Hydrogen can displace unreactive metals from their compounds

Copper is a fairly unreactive metal. It will leave copper compounds, such as copper (II) oxide, quite readily. Hydrogen can displace copper from copper (II) oxide. Figure 71.1 shows the apparatus which can be used to perform this reaction. Hydrogen is passed over heated copper (II) oxide. After a while, the black copper (II) oxide turns to pure, orange-pink copper.

The equation for the reaction is:

copper (II) oxide + hydrogen → copper + water
$$CuO(s) + H_2(g) \rightarrow Cu(s) + H_2O(g)$$

This is a displacement reaction. Copper is displaced by hydrogen. So hydrogen must be higher up the reactivity series than copper.

This reaction also works for tin and lead. Both tin and lead are below hydrogen in the reactivity series.

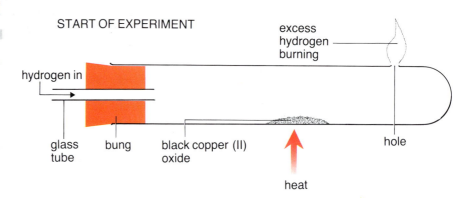

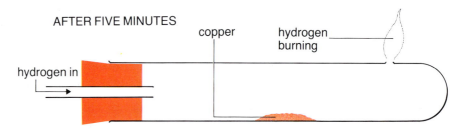

Fig. 71.1 Apparatus for the displacement of copper from copper (II) oxide by hydrogen. (Figure 86.1 shows a photograph of this experiment.)

Displacement is a way of extracting unreactive metals from their ores

The reaction between hot copper (II) oxide and hydrogen is an example of the type of reaction by which a metal can be extracted from its ore. The metal ore – in this case represented by copper (II) oxide – is heated with hydrogen, or with a metal higher in the reactivity series than the metal in the ore. The metal is displaced from its ore.

As the metal is extracted from its ore, it is **reduced**. Look again at the example of copper (II) oxide and hydrogen.

$$CuO(s) + H_2(g) \rightarrow Cu(s) + H_2O(g)$$
$$Cu^{2+} + 2e^- \rightarrow Cu \quad \text{Copper is reduced.}$$

This is, in fact, the origin of this sense of the word 'reduced'. Early metal workers noticed that the mass of ore they began with was always greater than the mass of metal they ended up with. So they said that the ore had been 'reduced'. We now have a more specific meaning for the word. The metal in the ore is said to be reduced because it gains electrons. The extraction of a metal from its ore is called **reduction**.

Fig. 71.2 The Egyptians are thought to have used a furnace like this to extract copper from copper ore around 1200 BC.

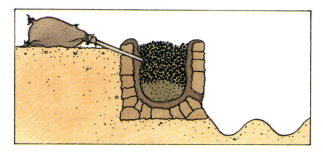

The furnace was built of stones and lined with clay. The copper ore was mixed with charcoal in the furnace. Air was pumped into the furnace, using goatskin bellows, making the charcoal burn fiercely.

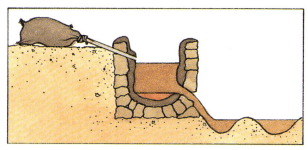

The carbon in the charcoal displaced the copper from its ore. The molten copper settled at the bottom of the furnace. The lighter slag was allowed to drain off.

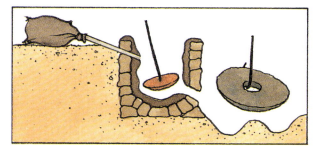

Carbon will displace unreactive metals from their ores

How did early humans reduce the ores of unreactive metals like copper and tin? They had no supply of hydrogen! But they did, unknowingly, use a displacement reaction.

When wood burns in a fire, it forms charcoal if the fire is short of air. Charcoal is almost pure carbon. Carbon is further up the reactivity series than copper and tin. So carbon will displace these metals from their ores. For example:

carbon + copper (II) oxide → copper + carbon dioxide
$$C(s) + 2CuO(s) \rightarrow 2Cu(s) + CO_2(g)$$

Copper ore was heated with charcoal, to extract the copper from its ore.

Carbon was also used to reduce iron ore, but a hotter furnace was required. This was achieved by mixing the ore with charcoal, lighting the charcoal, then using a bellows to pump air into the base of the furnace. This greatly increased the temperature of the furnace, so the rest of the charcoal was able to reduce the iron ore.

This method of ore reduction, **smelting**, works for all metals below carbon in the reactivity series. These include zinc, iron, lead, tin and copper. But it will not work for more reactive metals, such as aluminium. The ores of these more reactive metals are often reduced by melting them, and then passing electricity through them. But even this does not work for titanium ores, because titanium is extracted from titanium chloride, which is a covalent compound and so will not conduct electricity. Instead, titanium is displaced from its ore by a more reactive metal, magnesium.

titanium chloride + magnesium → titanium + magnesium chloride
$$TiCl_4(l) + 2Mg(s) \rightarrow Ti(s) + 2MgCl_2(s)$$

Fig. 71.3 Titanium alloy was used to make Concorde's fuselage.

Questions

1 a Write a word equation for the reaction of lead (II) oxide, PbO, with hydrogen.

b Write a balanced chemical equation for this reaction.

c Which substance is oxidised in this reaction?

d Which substance is reduced in this reaction?

2 How are the following metals extracted from their ores?

a zinc

b calcium

c sodium

d copper

e titanium

3 a Write a word equation for the reaction of lead (II) oxide with carbon.

b What properties of lead and carbon make this reaction a possible way of extracting lead from its ore?

Iron is the most widely-used metal

Iron is the fourth most abundant element in the Earth's crust. Aluminium is the only metal which is more abundant. The amount of iron used today is greater than the total amount of all other metals used. This valuable substance comes from various ores. These include **magnetite**, which contains Fe_3O_4, **haematite**, which contains Fe_2O_3, and **siderite**, which contains $FeCO_3$. We shall look at how iron is extracted from haematite.

Iron is extracted from its ore in a blast furnace

Iron ore goes into the blast furnace at the top. **Coke**, which consists almost entirely of carbon, and **limestone**, $CaCO_3$, go in with the ore. The iron ore, coke and limestone are called the **charge**.

Hot air is blasted into the base of the furnace through pipes called 'tuyeres' (pronounced 'tweers'). This hot air enables the coke in the bottom part of the furnace to burn:

carbon + oxygen → carbon dioxide
$$C(s) + O_2(g) \rightarrow CO_2(g)$$

This reaction heats the furnace, and produces carbon dioxide gas. The carbon dioxide rises up the furnace. Here, it reacts with more coke, forming carbon monoxide:

carbon + carbon dioxide → carbon monoxide

$$C(s) + CO_2(g) \rightarrow 2CO(g)$$

The carbon monoxide meets and reacts with the iron ore. It reduces the iron ore to iron:

$$Fe_2O_3(s) + 3CO(g) \rightarrow 2Fe(l) + 3CO_2(g)$$

Some of the iron ore reacts with the coke, which also reduces the iron ore:

$$2Fe_2O_3(s) + 3C(s) \rightarrow 4Fe(l) + 3CO_2(g)$$

The iron produced in these reactions falls to the base of the furnace. It forms a pool of molten iron.

This process is a continual one. The blast furnace runs for months at a time. New charge is constantly added.

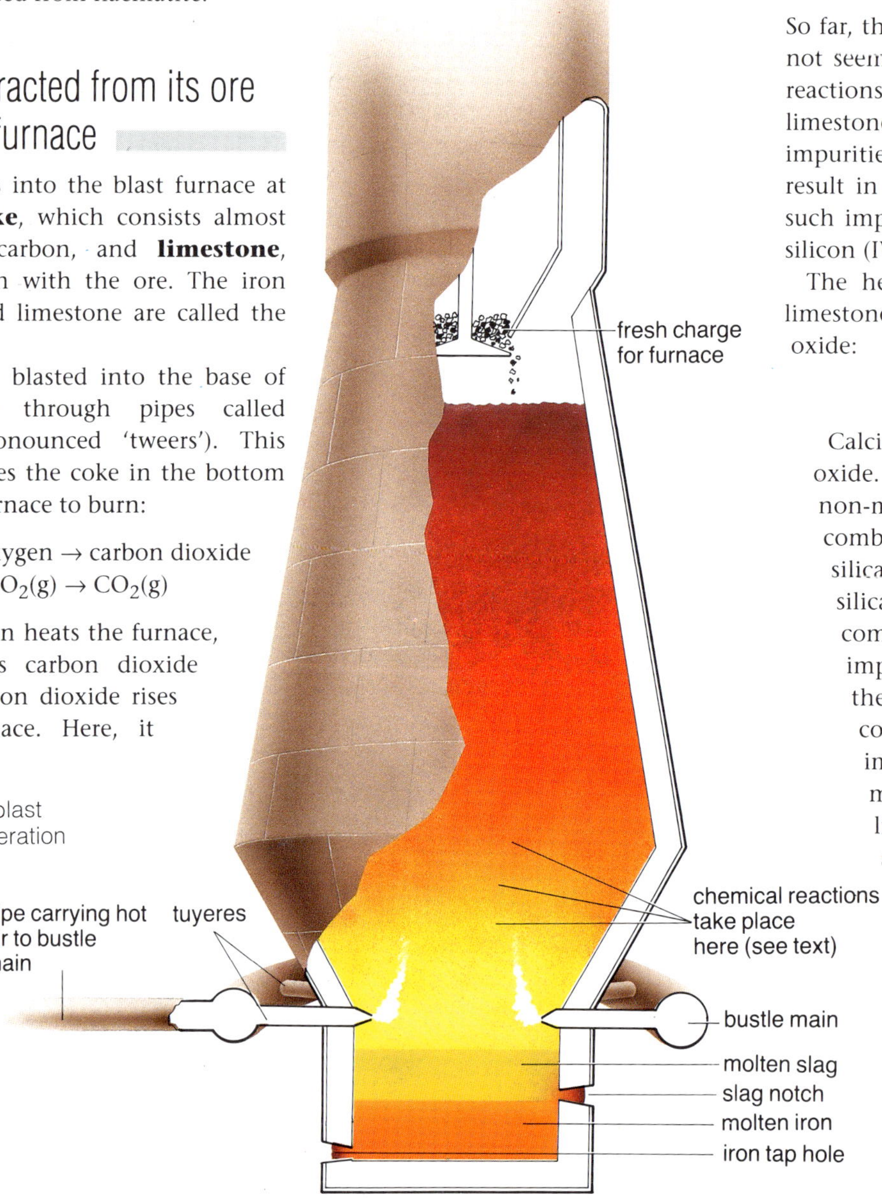

The limestone removes impurities

So far, the limestone in the charge does not seem to be involved in any of the reactions. So why is it added? The limestone acts as a **flux**, removing impurities which would otherwise result in a poorer quality of iron. One such impurity is sand, which is mostly silicon (IV) oxide, SiO_2.

The heat of the furnace causes the limestone to decompose to calcium oxide:

$$CaCO_3 \rightarrow CaO + CO_2$$

Calcium oxide is a *basic* metal oxide. Silicon (IV) oxide is an *acidic* non-metal oxide. The two oxides combine to form the salt calcium silicate, $CaSiO_3$. The calcium silicate, along with similar compounds formed from other impurities, falls to the bottom of the furnace. Here, the compounds formed from the impurities float on top of the molten iron. They look a bit like molten lava, and are called **slag.**

Every now and then the slag has to be run out of the furnace, through a hole called a **tap hole** or **slag notch**. The molten slag solidifies as it cools. It is broken up and used for road building. Other uses are now being developed for what used to be just a waste product.

Fig. 72.1 A blast furnace in operation

pipe carrying hot air to bustle main

tuyeres

fresh charge for furnace

chemical reactions take place here (see text)

bustle main

molten slag

slag notch

molten iron

iron tap hole

The iron is used for making steel

The molten iron is run out of the furnace, through another tap hole, into waiting containers. It may be allowed to set in a mould, in which case it is called **pig iron** or **cast iron**. Cast iron is not very useful. It contains about 4% carbon, either dissolved in, or combined with, the iron. Cast iron also contains other impurities. This gives the iron a poor combination of properties, including brittleness.

Most of the iron is therefore used to make **steel**. This involves removing most, but not all, of the carbon. The other impurities are also removed. Other metals may then be mixed in with the steel. Steel is therefore an alloy. It contains iron, a little carbon, and sometimes other metals too.

Steel-making requires molten cast iron, so steel-making plants are often very near to blast furnaces. Otherwise, the cast iron would solidify, and would then have to be reheated for steel-making.

Different substances mixed with the iron make different types of steel

Before the steel is made, the molten cast iron is first analysed to find out exactly how much carbon and other impurities it contains. The amount of carbon is lowered by blowing oxygen through the molten metal. The oxygen combines with the carbon, forming carbon dioxide, which escapes as a gas. The amount of oxygen used is calculated, to lower the carbon content to the level which the steel-makers want. For a malleable mild steel, such as is used for making car body panels or food cans, the carbon level is about 0.2%. For tougher, more resilient steel, the ideal carbon level is about 0.5%.

Some of the other impurities, like silicon, also form oxides in the steel-making ladle, so lime, CaO, is added. This combines with these oxides to form a slag of calcium silicate and other compounds. The slag can then be scraped off.

Alloying with other metals produces specialist steels. **Stainless steel** is 70% iron, 20% chromium and 10% nickel. This is a highly corrosion-resistant, but rather expensive type of steel. It is used for making cutlery. For cutting tools which have to maintain a sharp edge, even when hot, **high speed steel** is used. This contains tungsten and sometimes chromium mixed in with the iron. Concorde needed to be made of a metal which was strong yet light, and which would keep its properties even when heated up by air friction. The answer was a steel alloy containing a lot of titanium. Other metals which may be added to iron include molybdenum and manganese. Both of these bring about a great increase in hardness.

Fig. 72.2 Molten slag is poured off from a blast furnace into containers.

Fig. 72.3 Molten iron runs into a ladle from a blast furnace. It will be poured into casting moulds.

Fig. 72.4 A steel-making furnace

Questions

1 a The coke added to a blast furnace fulfills two functions. What are they?

b Why is limestone added to a blast furnace?

c The exhaust gases from a blast furnace contain:
  i nitrogen
  ii carbon dioxide
  iii carbon monoxide.
  Explain the presence of each of these gases.

d The exhaust gases are hot. What use could be made of this heat by the iron-producing plant?

2 What effect does lowering the carbon content have on cast iron?

3 Give two reasons why people are trying to find uses for slag.

4 a During steel making, the steel must be kept molten for as long as possible. Heating, however, is very rarely needed. Why is this?

b Why is accurate analysis of the molten cast iron necessary?

5 Find out about the roles played by the following in steel making:
  a water cooled lances
  b magnesium
  c argon
  d scrap.

6 Use reference books to find out about the parts the following people played in the development of the modern steel industry:
  a Henry Bessemer
  b Andrew Carnegie.

EXTENSION

Ionic compounds conduct electricity when molten or dissolved in water

If the substances that conduct electric current are investigated, there seems to be a simple rule. Metals conduct and non-metals do not – with the exception of carbon. However, there is a type of substance which will conduct electricity in some states but not in others. An example is sodium chloride.

Figure 73.1a shows sodium chloride behaving as a typical non-conductive non-metallic substance. But if the sodium chloride is heated enough to melt it, as in Figure 73.1b, a current begins to flow. Molten sodium chloride conducts an electric current well.

Sodium chloride dissolved in water will also conduct an electric current, despite the fact that neither sodium chloride nor water are good conductors when pure and separate from each other.

Many other substances behave in the same way, only conducting electricity when they are molten or dissolved in water. They include potassium bromide, sodium hydroxide, copper sulphate and zinc nitrate. All these substances are **ionic compounds**.

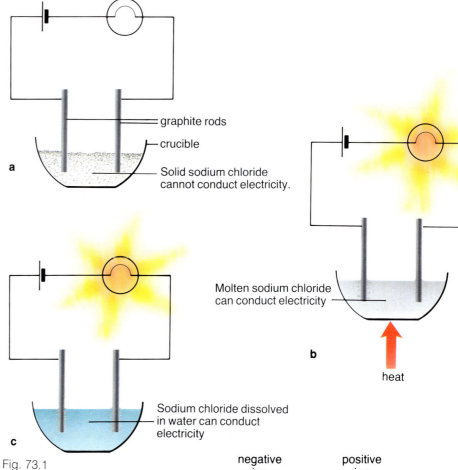

graphite rods

crucible

a Solid sodium chloride cannot conduct electricity.

Molten sodium chloride can conduct electricity

b

heat

Sodium chloride dissolved in water can conduct electricity

c

Fig. 73.1

Ions are attracted to positive and negative electrodes

How do molten or dissolved ionic substances conduct electricity? They do not have electrons which are free to move, as metals do, so the current cannot be carried as a drift of electrons. But they do contain positive and negative **ions**. Sodium chloride, for example, contains Na^+ and Cl^- ions. The movement of these ions makes up the electric current.

In Figure 73.2, the two graphite rods connected to the cell are called **electrodes**. The rod attached to the positive side of the cell becomes positively charged. It becomes a positive electrode, or **anode**. It attracts the negatively charged chloride ions, which move towards it.

The graphite rod attached to the negative side of the cell becomes negatively charged. It becomes a negative electrode, or **cathode**. It attracts the positively charged sodium ions, which move towards it.

This movement of charged particles between the electrodes is an electric current. An electric current flows between the electrodes, so the circuit is complete and the bulb lights.

Why won't solid sodium chloride conduct electricity? It does consist of Na^+ and Cl^- ions, but in the solid they are bound in one place. They are not free to move and so cannot make up an electric circuit. Molten sodium chloride, or sodium chloride dissolved in water, can conduct electricity because the ions are free to move around.

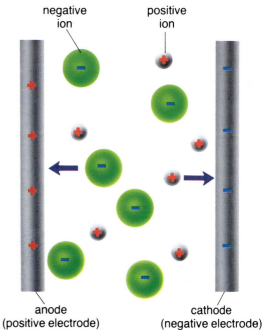

negative ion

positive ion

anode (positive electrode)

cathode (negative electrode)

Fig. 73.2   If positive and negative ions are free to move – as they are when in solution or in a liquid state – then negative ions go to the anode and positive ions go to the cathode.

Sodium chloride decomposes as a current flows through it

When an electric current flows through molten sodium chloride, bubbles are seen around the positive electrode. A choking yellow gas is produced. This gas is **chlorine**. At the negative electrode, molten **sodium metal** forms. The electric current is causing the sodium chloride to split up into the elements from which it is made.

$$2NaCl(l) \longrightarrow 2Na(l) + Cl_2(g)$$

This is called **electrolysis**. Electrolysis is the decomposition of a compound, caused by electricity.

Sodium chloride *solution* also undergoes electrolysis as a current flows through it, but different products are formed. At the positive electrode, bubbles of **chlorine** gas form as for molten sodium chloride. But at the negative electrode, **hydrogen** gas forms, and, although you cannot see it, **sodium hydroxide** forms in solution. The products are different because the water is also involved in the electrolysis reaction.

$$2NaCl(aq) + 2H_2O(l) \rightarrow H_2(g) + Cl_2(g) + 2NaOH(aq)$$

All ionic compounds undergo similar electrolysis reactions, producing new substances when an electric current passes through them, when molten or in aqueous solution. They are known as **electrolytes**.

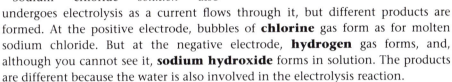

Fig. 73.3 Molten lead bromide conducts electricity. Electrolysis releases bromine gas and leaves lead in the crucible.

What forms at each electrode?

During an electrolysis reaction, a new substance forms at each electrode. How can you predict what these new substances will be?

For *molten electrolytes*, this is quite easy. The positive metal ion is attracted to the cathode, so the metal element forms here. The negative non-metal ion is attracted to the anode, so the non-metal element forms here. Table 73.1 gives some examples.

For *electrolytes in aqueous solution*, things are a little more complicated. The substance formed at the cathode is usually hydrogen, unless the solution contains copper or silver ions. In this case, copper or silver will form at the cathode. The substance formed at the anode is usually oxygen, unless the solution contains chloride, bromide or iodide ions. In this case, chlorine, bromine or iodine will form at the anode, and may dissolve in the solution. Table 73.2 gives some examples.

Questions

1 a Why does sodium chloride conduct electricity when molten or in aqueous solution, but not when solid?

b A covalent compound like sugar will not conduct when solid, molten, or in aqueous solution. Explain why this is so.

2 Explain the terms:
  a electrolysis
  b electrolyte.

3 Which of the following substances are:
  a conductors,
  b non-conductors,
  c electrolytes?
  A zinc chloride (aq) **B** mercury
  C sodium hydroxide (aq)
  D sulphur **E** iron **F** water

4 What forms at each electrode when the following are electrolysed?
  a KCl(l)   **b** NaBr(l)
  c MgO(l)   **d** PbI_2(l)
  e Ag_2SO_4(aq) **f** $Cu(NO_3)_2$(aq)
  g H_2SO_4(aq)   **h** $MgCl_2$(aq)

5 Electrolysis is used extensively in the chemical industry for making such things as sodium hydroxide solution, sodium and magnesium metals, and chlorine gas. Explain how each of these could be made.

Table 73.1   Products of electrolysis of molten electrolytes

Electrolyte	Positive ion	Substance at cathode	Negative ion	Substance at anode
Al_2O_3(l)	Al^{3+}	aluminium	O^{2-}	oxygen
$PbBr_2$(l)	Pb^{2+}	lead	Br^-	bromine
$CaCl_2$(l)	Ca^{2+}	calcium	Cl^-	chlorine

Table 73.2   Products of electrolysis of aqueous electrolytes

Electrolyte	Substance at cathode	Substance at anode
$MgSO_4$(aq)	hydrogen	oxygen
$Zn(NO_3)_2$(aq)	hydrogen	oxygen
KOH(aq)	hydrogen	oxygen
$CuSO_4$(aq)	copper	oxygen
$CaCl_2$(aq)	hydrogen	chlorine
$AgNO_3$(aq)	silver	oxygen
$CuCl_2$(aq)	copper	chlorine
H_2SO_4(aq)	hydrogen	oxygen

74 ALUMINIUM

Aluminium is a reactive metal, which is extracted from its ore by electrolysis.

Aluminium is a very abundant metal

Aluminium is the most plentiful metal in the Earth's crust. Of all the elements, only oxygen and silicon are more plentiful. Aluminium makes up 8% of the crust. It is found in the minerals bauxite, mica and cryolite, and in clay.

However, until about 100 years ago, aluminium was virtually a precious metal. Despite its abundance, it was very rare as a pure metal, because it was so hard to extract from its ore. This is because aluminium is a reactive metal. So it cannot be extracted by smelting with carbon. Displacement reactions were tried, but metals like sodium or potassium had to be used, resulting in too high a cost. Electrolysis of the molten ore was tried, but the most plentiful ore, bauxite, contains aluminium oxide, which does not melt until it reaches 2050 °C.

Fig. 74.1   A bauxite mine in Jamaica

Aluminium is extracted by the Hall-Héroult process

The solution to the problem of extracting aluminium from its ore was discovered by Charles Hall in the USA and by Paul Héroult in France, working completely independently. The method we now use for extracting aluminium from its ore is called the Hall-Héroult process.

1 Bauxite, which is impure aluminium oxide, is purified. This gives Al_2O_3.

2 The aluminium oxide is dissolved in molten cryolite, Na_3AlF_6, at a temperature of between 800 and 1000 °C. This is done in a steel vessel.

3 The mixture is electrolysed. This involves lowering graphite blocks into the molten mixture. The graphite blocks are connected to the positive side of a large electrical power supply, making them **anodes**. The steel vessel is connected to the negative side of the power supply, making it the **cathode**.

In the molten mixture, the Al^{3+} and O^{2-} ions from the Al_2O_3 are free to move about. The O^{2-} ions, being negative, are attracted to the positive anodes. Here they lose electrons and oxygen gas forms:

$$2O^{2-} - 4e^- \longrightarrow O_2$$

Due to the high temperatures the oxygen reacts with the graphite anodes, which slowly burn away and need to be replaced periodically.

The Al^{3+} ions, being positive, are attracted to the negative cathode. Here, they gain electrons and liquid aluminium forms:

$$Al^{3+} + 3e^- \rightarrow Al$$

So aluminium is smelted from its ore. The process is not cheap, but aluminium can now be produced cheaply enough to make it the second most widely used metal. The main cost of the process is the electrical energy consumed. So much electricity is required that aluminium smelters have their own power stations nearby.

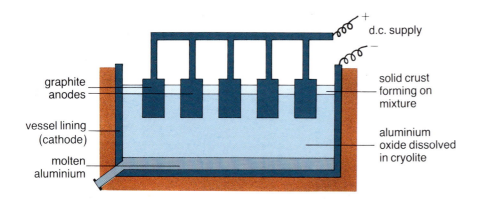

graphite anodes

vessel lining (cathode)

molten aluminium

d.c. supply

solid crust forming on mixture

aluminium oxide dissolved in cryolite

74.2   The Hall-Héroult process

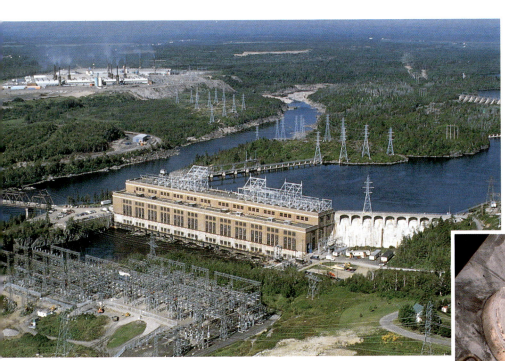

Fig. 74.3 The hydroelectric power station shown here provides electrical energy for the aluminium smelting plant in the background.

Aluminium becomes covered with a film of oxide

Aluminium is a reactive metal, which is why it is so difficult to extract from its ore. So why is it such an easy metal to handle and use? How can we wrap a chicken in aluminium foil and put it in the oven, without the foil burning in the hot oven? Why does the aluminium not react with the food?

Aluminium metal reacts with oxygen in the air to form a film of aluminium oxide on its surface. This film covers the aluminium completely. It is very unreactive, so the aluminium behaves as if it were unreactive. If aluminium foil is put into hot hydrochloric acid, at first nothing seems to happen. But as soon as the acid has dissolved away the oxide film, a very vigorous reaction takes place.

If additional resistance to corrosion is needed, the aluminium can be **anodised**. This increases the thickness of the oxide film. The thicker oxide film on anodised aluminium also takes up dyes very well, making it possible to dye the aluminium attractive colours.

As well as resistance to corrosion, aluminium has other very useful properties.

- It has a low density.
- It can be easily rolled into thin sheets.
- It has good electrical conductivity.
- It has good thermal conductivity.

These properties have led to its wide range of uses.

Fig. 74.4 Pouring molten aluminium

Fig. 74.5 Aluminium is used for making cables.

Questions

1 Bauxite is impure aluminium oxide. A common impurity is iron oxide, Fe_2O_3. Aluminium oxide dissolves in alkali, but the iron oxide does not. How might the bauxite be purified?

2 a Explain the significance of the oxide film which forms on the surface of an aluminium object.

 b Why is aluminium, and not iron, used for making cooking foil?

3 a When aluminium is extracted from aluminium oxide, the aluminium is reduced. Give two definitions of reduction which fit this statement.

 b What is oxidised in this process?

EXTENSION

Sodium is extracted from salt

Sodium is a highly unusual metal. Of the metals you have already met, it is most similar to lithium. It is soft, has a density so low that it floats on water, and is very reactive.

Sodium is a very common element, making up 2.6% of the Earth's crust, always in combination with other elements. Its most important ore is **salt** or **sodium chloride**, NaCl. This may be obtained from sea water, or mined as rock salt. The main rock salt mines in Britain are in Cheshire. Great underground caverns are produced, as the salt is first loosened by blasting, and then removed in trucks or by conveyor, or pumped out as saturated salt solution.

Since sodium has a reactivity similar to that of lithium, it is difficult to extract it from its ore. Smelting with carbon will not work, because sodium is much more reactive than carbon. Like aluminium, it has to be extracted by electrolysis.

The cell used is called a **Down's cell**. The sodium chloride has to be melted so that it will conduct electricity. The chloride ions are attracted to the anode, where they lose electrons:

$$2Cl^- - 2e^- \rightarrow Cl_2$$

Chlorine gas is formed. The chloride ions are oxidised, as they lose electrons.

The sodium ions are attracted to the cathode, where each ion gains an electron:

$$Na^+ + e^- \rightarrow Na$$

Molten sodium forms. The sodium ions are reduced, as they gain electrons.

It is important to keep the liquid sodium separate from the chlorine gas, or they would react vigorously to reform sodium chloride.

Fig. 75.1 Sodium chloride occurs naturally as rock salt. These crystals are growing in a salt water spring in Iran.

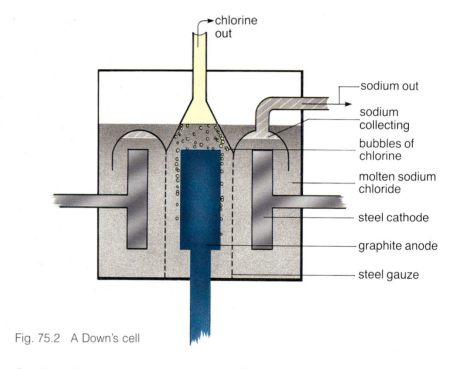

chlorine out

sodium out

sodium collecting

bubbles of chlorine

molten sodium chloride

steel cathode

graphite anode

steel gauze

Fig. 75.2 A Down's cell

Sodium is used as a coolant and in street lamps

Because of its reactivity, sodium has few of the uses normally associated with metals. It reacts vigorously with water, so it can have no uses where it might get wet. However, its high reactivity does mean that it can be used to extract titanium from titanium chloride:

$$TiCl_4 + 4Na \rightarrow Ti + 4NaCl$$

Magnesium, however is more commonly used than sodium for this reaction.

Sodium can be used as a **coolant** in nuclear power stations. It takes heat from the hot nuclear reactor to the boilers, where it is used to boil water. It is suitable for this purpose because it conducts heat well, flows easily when melted, and is liquid over a very large temperature range. Sodium melts at 98°C and does not boil until it reaches 891°C. But care must be taken to keep the hot liquid sodium away from air and moisture.

Yellow street lamps contain sodium vapour. Sodium vapour gives out a lot of visible light energy per joule of electrical energy transferred, so these lamps are cheap to run.

Chlorine compounds can be useful, but can also damage the environment

The Down's cell produces chlorine as well as sodium. The element chlorine is a poisonous yellow gas, heavier than air. One of its uses is in water treatment; it is used to kill bacteria in tap water and swimming pools. It is also used in the manufacture of a variety of products including:

- anaesthetics
- pesticides, including DDT
- solvents
- plastics like PVC
- bleaches
- CFCs, which can be used as the coolant in refrigerators and as the propellant in aerosols.

You will recognise many of these as being bad for the environment. The use of CFCs is being reduced, now that their role in damaging the ozone layer is understood. DDT kills harmful insects, but also useful ones, and is also damaging to small animals like mice and birds, which eat crops sprayed with DDT. Predatory birds such as owls and peregrine falcons may eat small animals with DDT in their bodies, and are then badly affected themselves. The use of DDT is now strictly regulated.

Sodium hydroxide and hydrochloric acid production

Another way of producing chlorine from sodium chloride is by using a **flowing mercury cell**. Here, the sodium chloride is dissolved in water, rather than being melted. As in the Down's cell, chlorine forms at the anode and sodium at the cathode. But in this cell, the cathode is mercury. The sodium dissolves in the liquid mercury.

The sodium/mercury mixture flows out of the cell, and into water. The sodium reacts with the water, forming **sodium hydroxide**:

$$2Na + 2H_2O \rightarrow 2NaOH + H_2$$

Sodium hydroxide is a strong alkali. It is used in the manufacture of paper, soap and bleaches, and in many important stages in the manufacture of other products of the chemical industry. It is also used to purify bauxite, the main ore of aluminium, producing pure aluminium oxide.

The hydrogen produced in this reaction can be made to react with the chlorine formed in the cell. **Hydrogen chloride** is formed, which produces **hydrochloric acid** when dissolved in water. This is an important industrial acid.

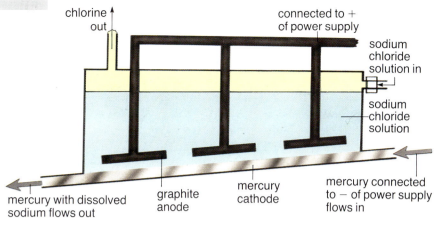

Fig. 75.3   A flowing mercury cell

Fig. 75.4   Chlorine, and gases made from chlorine, have been used in chemical warfare. They have a choking, blinding effect, and frequently cause permanent lung damage.

Questions

1 a Why is sodium hard to extract from its ore?

b Why is sodium not used to make metal objects?

c Would you expect sodium to 'rust' quickly? Explain your answer.

2 Compare the Down's cell with the flowing mercury cell, by answering the following questions.

a What does the Down's cell produce which the flowing mercury cell does not?

b What additional pollution problems does the flowing mercury cell involve?

c Which process is likely to be cheaper? Why?

---

3 PCBs are chlorine compounds. Find out what PCB stands for, what PCBs are used for, and why these compounds are so dangerous.

└ *EXTENSION* ─────────

76

MAKING METALS LAST

Supplies of metals are not limitless. We need to use metals sparingly, if they are not to run out in the future.

Metals are a finite resource

Without metals, we would be back in the Stone Age. But we do not have limitless supplies of metals. If we keep drawing on this finite resource, it will eventually run out. But, with intelligent control of our uses of metals, and treatment of them during and after use, we can extend their time of availability.

Metals can be replaced by other substances

Many of the roles traditionally filled by metals can also be filled by other substances. As metal supplies dwindle, they become more expensive. So manufacturers have economic reasons to look for alternatives.

There are some uses for which metals are still the only choice. Where electrical conductivity is required for everyday use, then metals must be used. Metals are still the best choice when both strength and heat resistance are needed. But car manufacturers, for example, are conducting extensive research into the use of plastics and ceramics for engine parts. Fibreglass-reinforced plastic is already used for car body panels. Saucepans can be made from heat-resistant glass, dustbins from plastic, and roofing from PVC instead of lead.

The amounts of metal used can be reduced

If metal must still be used for a particular purpose, then it may be possible to use less. This is often made possible by using alloys instead of pure metals. In aircraft construction, where lightness is essential, carefully chosen alloys can provide all the necessary strength and durability, using less metal than would be needed otherwise. Drills and other tools, which use the hard metal tungsten, often only have tungsten at their tips; the rest of the tool is made of cheaper and more abundant metals.

Preventing corrosion extends the life of metals

We use far more iron than any other metal. Iron is strong, and comparatively cheap. But unprotected iron rusts. It slowly combines with oxygen to form iron oxide, Fe_2O_3, which is rust. The rust crumbles away, allowing the iron underneath to come into contact with air and rust in its turn.

Rust is a weak, crumbly material. As a piece of iron turns to rust, it loses all its strength. It becomes unattractive to look at. By preventing rusting, the lifetime of an iron or steel article is greatly increased. So the amount of iron we use is decreased.

Fig. 76.1 Using notes instead of coins reduces our use of metals. Some bank notes are now made of plastic, which lasts much longer than paper. These Australian notes are made of polypropene.

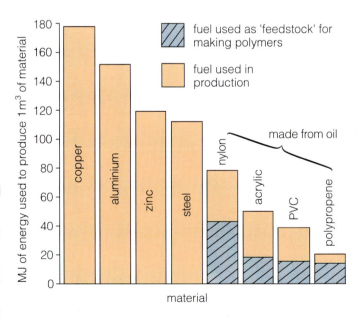

Fig. 76.2 Obtaining metals and other materials requires energy. If we make the most of the metals we obtain, we can keep energy use down. This not only helps to conserve our limited resources, but also decreases pollution.

Do air and water affect the rate at which iron rusts?

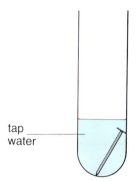
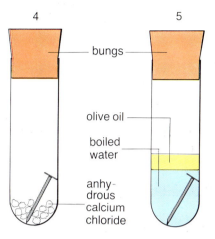

Fig. 76.3

1 Number five dry boiling tubes one to five.

2 Thoroughly clean five iron nails with an emery cloth, so that they are completely rust free. Place one nail in each boiling tube.

3 Leave just the nail in tube 1.

Place a wad of wet cotton wool in tube 2.

Pour water into tube 3 to cover the nail.

Put a drying agent into tube 4. This could be anhydrous calcium chloride, or silica gel, which will remove water vapour from the air. Put a bung in the tube.

Pour water into tube 5 to cover the

nail, then add about 2 cm more. Add a layer of olive oil about 0.5 cm deep. Carefully boil the water for at least ten minutes, making sure that the liquids do not spit out of the tube. This will drive off any air dissolved in the water. Put a bung in the tube.

You could set up more tubes, with different conditions, if you like.

4 Copy the table below.

Tube	1	2	3	4	5
conditions					
after days					
after days					

5 Fill in the 'conditions' row. For tube 1, for example, you could write 'exposed to air, but very little water (only the water vapour in the air).'

6 Examine your tubes a day or two later, and record the amount of rusting. Continue to do this at intervals of two or three days.

Questions

1 Rank the five nails in order of speed of rusting.

2 What are the most corrosive conditions for iron?

3 Do either or both air and water cause iron to rust?

Questions

1 a Why is it difficult to replace metals with other materials?

b What are the advantages and disadvantages of replacing:

i an aluminium saucepan with a glass one?

ii a lead roof covering with a PVC membrane?

iii a mild steel car bonnet with a plastic one?

c Give some more examples of metal replacement.

2 The following are made of metal:

i long span bridges

ii electrical wiring

iii coins.

We could save on metal resources by using less metal in each situation, or by using a non-metal instead. Discuss the problems that would be involved in each situation.

3 a Why is rusting a threat to metal resources?

b What chemical substance is iron rust?

c How are the rusting of iron and the 'rusting' of aluminium different?

d What factors cause iron to rust?

77 PREVENTING CORROSION

Iron, and most types of steel, rust very easily. Various methods are used to reduce the rate of this corrosion.

Painting or oiling can protect iron and steel

Iron is made useless by the formation of iron oxide, or rust. This process is greatly speeded up if the metal is in contact with both air and moisture. If the metal is coated with something which protects it from these substances, it will not rust.

Painting is a cheap and simple way of protecting iron. Car bodies, ships' hulls, bridges and iron railings are all protected by layers of paint. But paint may flake or chip off, so care of paintwork and repainting is essential if rusting is not to occur in damaged areas.

In some cases, a layer of **oil** is better than a layer of paint. An oil film on the surface of an iron or steel object will provide a barrier against air and moisture. This can be used where paint would easily get chipped, for example on tools. But the oil must be constantly reapplied, so this method is unsuitable for large structures.

Fig. 77.1 This steel lamp post has been galvanised to protect the steel from rusting.

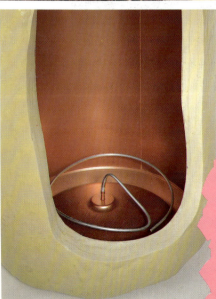

Fig. 77.2 Hot water and heating systems contain many different metals. Aluminium is more reactive than the iron, copper and brass in the system. So an aluminium rod in the bottom of the water tank protects these other metals from corrosion. The aluminium rod is a sacrificial anode.

DID YOU KNOW?

Most car manufacturers use many methods to reduce corrosion on the cars they make. They may use more than 10 kg of wax, and 22 coats of paint to protect a metal car body. The body may also be galvanised. Exhaust systems may be made of stainless steel, which costs three times as much as mild steel but which lasts far longer.

Iron and steel can be coated with other metals

Zinc or **tin** can be used to coat iron to stop it from rusting. Neither method is cheap, but they both provide good protection for the iron or steel.

Tin plate is made by thoroughly cleaning a sheet of steel, and then dipping it into molten tin. A layer of tin only 0.001 mm thick coats the steel. This is how food 'tins' are made. They are really made of steel, with just a very, very thin layer of tin on their surfaces. The tin coating stops air and water from coming into contact with the iron. It also protects the iron from acids in the canned food. This not only stops the iron being corroded, but also stops iron salts from contaminating the food inside the tin. Tin is much less reactive than iron, and is not attacked by the acids in food.

When two metals are in contact with one another, and in contact with air and moisture, only the more reactive one will rust. Iron is more reactive than tin, so if the tin coating is damaged, the iron will rust. It will rust even more rapidly than when iron is on its own. So tin plate is only usable in situations where it will not get knocked or scratched.

Steel or iron can also be coated with **zinc**. This is called **galvanising**. Zinc is a more reactive metal than iron, so if galvanised steel is scratched, the zinc corrodes, but not the steel. However, the zinc corrodes only slowly. Galvanised steel is used for buckets, car body panels and girders, especially if these are exposed to corrosive situations.

The steel does not need to be completely covered by zinc. If blocks of zinc are bolted to a sheet of steel, the zinc will corrode instead of the steel. This is called **sacrificial corrosion**. It is used to protect ships' hulls. It is important to make sure that the zinc blocks are replaced before they have completely corroded away.

Alloying can reduce rusting

In 1913, the British metallurgist Harry Brearley was making various different steel alloys, looking for one suitable for making gun barrels. He threw his failures out onto a rubbish heap. Several months later, he noticed that among the pile of now rusty scrap one piece remained bright and uncorroded. He found, on analysing it, that it was an alloy containing 14% chromium. He had produced **stainless steel**, an iron alloy which can be left exposed to air indefinitely without corroding.

In fact, stainless steel does corrode, but, like aluminium, a tough layer of oxide forms on its surface, preventing further corrosion.

Modern stainless steels consist of approximately 70% iron, 20% chromium and 10% nickel, with a trace of carbon. Alloying is the most expensive form of corrosion protection, but it is also the most complete, because no maintenance is needed.

Fig. 77.3 Salt is added to roads in winter to lower the freezing point of the water on the road surface, and reduce the chance of ice forming. But the salt increases the rate of corrosion of bicycles and cars. How could this corrosion be reduced?

INVESTIGATION 77.1

Comparing different methods for preventing corrosion

You are going to look at steel which has been protected in various ways.

1 Take pieces of each type of steel; for the samples of coated steels, take two pieces one of which has had its protective coating damaged. Put each piece of steel into a test tube containing a little tap water.

2 Leave the steel pieces for several days, and then examine them for traces of rust on the metal or in the water. Record your observations.

3 Record your results and draw conclusions, comparing in particular the rates of rusting of the damaged samples.

Why not try:

- kicking a sample around outside for a few minutes, to see how tough the coatings are?
- putting drops of various acids onto the samples?
- winding magnesium ribbon around a nail before putting it into tap water?
- leaving some samples in contact with some tinned fruit?
- any other tests you can think of?

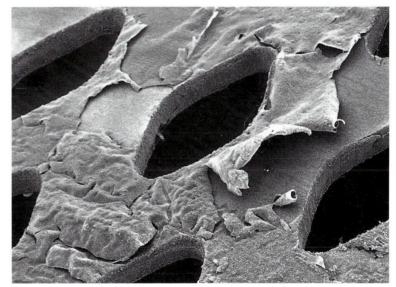

Fig. 77.4 This scanning electron micrograph shows the surface of a silver mesh solar cell interconnector from the Space Shuttle. The shuttle orbits at altitudes of between 250 and 300 km. Here, the Earth's atmosphere is very thin, containing only about 10^9 particles per cm^3. But some of these particles are oxygen atoms, and the shuttle flies through them at 8 km/s. Oxygen atoms are much more reactive than oxygen molecules. The oxygen atoms react with the surface of the shuttle. In this photograph, flakes of silver oxide can be seen, which formed during an eight day shuttle flight. This sort of corrosion poses serious problems. Space engineers are working to find solutions to these problems, such as developing new protective coatings for shuttle components.

Questions

1 Which type of anti-corrosion treatment is used for:
  a food cans?   **e** scalpels?
  b garden shears? **f** ships?
  c kitchen knives? **g** car bodies?
  d girders?   **h** bridges?

2 a List the main ways of preventing corrosion.
  b Which methods for preventing corrosion are easiest to damage?
  c With which method does damage result in accelerated corrosion?
  d With which method does damage result in least corrosion?

3 What is sacrificial corrosion? Why is this a suitable name? In terms of sacrifical corrosion, explain what happens to:
  a scratched tin plate.
  b scratched galvanised steel.

4 Find out what is meant by 'cathodic protection'.

─ *EXTENSION* ─

Safety must be considered when choosing metals

In Topic 11, you may have noticed how metals and their alloys are suited to their different tasks because of their different properties. But it is not only physical properties, such as hardness and strength, or chemical properties such as resistance to corrosion, which are important. Cost, availability and safety must also be considered.

Magnesium, for example, has a very low density, which could make it a useful light-weight structural metal. But it is very flammable. Mercury is unique as a liquid metal, but it is highly toxic. These negative safety aspects put considerable limits on the uses of these two metals.

Aluminium alloys are widely used where lightness is needed

Aluminium alloys combine strength and lightness. The density of aluminium is $2.70\,\text{g/cm}^3$, compared with $7.86\,\text{g/cm}^3$ for iron and $8.92\,\text{g/cm}^3$ for copper. In addition aluminium does not corrode.

But pure aluminium is neither very hard nor very strong. Alloying it with other metals can produce a hard, strong material, with aluminium's lightness.

The alloy **Duralumin** contains approximately 4% copper and 1% magnesium mixed with aluminium. Its density is nearly as low as that of pure aluminium, but it is much stronger and harder. It is used for aircraft bodies. Where even greater strength and hardness is required, **aluminium bronze** may be used. This is a mixture of approximately 90% copper and 10% aluminium. It has a high resistance to corrosion, and is used for ship's propellors and other items used in highly corrosive environments. But as ~~minium~~ bronze contains such a high ~~p~~ortion of copper, it has a fairly ~~d~~ensity, around $7.5\,\text{g/cm}^3$.

Fig. 78.1 Magnesium alloy is used for making bicycle frames, as it is an excellent light-weight structural material. Pure magnesium could not be used because it is too flammable.

Fig. 78.2 A piece of uranium-235, a radioactive isotope which is used as a fuel in nuclear reactors.

Fig. 78.3 The lightness, appearance and resistance to corrosion of this aluminium alloy make it an attractive building material.

Questions

1 Read the following passage, and then answer the questions which follow.

Lead

Lead has many of the same properties as gold. It is easy to work, and resists corrosion. Because of this, lead used to be used for making water pipes. The Latin word for lead is 'plumbum', so people who connected these pipes were called 'plumbers'.

It is now known that lead is toxic, causing brain damage, especially in children. Lead is also an expensive metal. So water pipes are now more likely to be made of copper or plastic.

Another important use of lead was in making sulphuric acid. Lead sheets were melted together to form huge boxes in which sulphuric acid could be made and kept. Lead melts at quite a low temperature, so it was not difficult to form the boxes, and any holes could be mended simply by melting the lead around the damaged area and letting it resolidify. Also, lead is not corroded by sulphuric acid.

Lead is no longer used in sulphuric acid plants, because the modern process operates at temperatures that would melt lead.

But lead is far from being a thing of the past. It is still used for roofing, and particularly for small pieces of 'flashing' to keep water from leaking through in awkward places like the base of a chimney. Again, it is the mouldability and resistance to corrosion of lead that make it suitable for these purposes.

Lead blocks ionising radiation. Radiographers wear lead aprons to protect themselves from the X rays they use. If they did not, they might suffer damage after being exposed to these rays day after day. Glass can be made using the compound lead oxide. This glass can then be used to make windows which enable scientists to look into nuclear reactors without the dangerous radiation getting out.

---

a Explain why lead's ease of working and resistance to corrosion once made it suitable for some of the uses mentioned.

b Take one of your answers to part a, and explain why lead is no longer used for this purpose.

c Explain the use of lead in the handling of radioactivity.

d What advantage does lead glass have over lead sheet?

2 Woods metal is an alloy of lead, tin, cadmium and bismuth. It melts at 71°C.

a What would you see if Woods metal was put into boiling water?

b Why is Woods metal not used as a solder?

c How does the automatic sprinkler in the diagram below turn itself on during a fire?

d Woods metal is used in automatic fire alarms. The diagram below shows a simple automatic alarm circuit containing a strip of Woods metal.

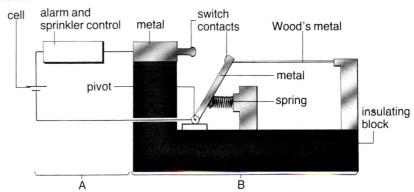

i Why is the alarm normally off?

ii In the event of a fire, describe the sequence of events which turns the fire alarm on.

3 Gold is often used to make alloys. Pure gold is known as 24 carat gold. An alloy containing 50% gold is called 12 carat gold. Alloying the gold makes it harder.

a If you wear a 24 carat gold ring next to a 12 carat gold ring, which one wears the other down?

b What percentage of gold is contained in 18 carat gold ?

c As well as producing extra hardness, what other advantage is there in alloying gold, rather than using the pure metal ?

4 The figures below give the energy costs of producing one tonne of seven commonly used materials.

material	MJ per tonne
aluminium	250 000
plastic	54 000
zinc	50 000
lead	50 000
iron and steel	47 000
copper	43 200
cement	7 250

a Draw a histogram to show these figures.

b Compare the figures for aluminium and copper with those in Figure 76.2. Can you explain the differences between the two graphs?

c Why is it so expensive to produce aluminium?

d Copper is cheaper to produce than iron or steel. Why do we choose to use steel for making cars, and not copper?

e Give one use of each of the materials listed in the table. In each case, give one reason why this material is chosen for this purpose.

---

f It takes 7 MJ of energy to make an aluminium drink can. It takes 321 MJ of energy to make a loaf of bread. Make a list of the stages of production of these two items, and suggest why the loaf of bread requires so much more energy.

EXTENSION

Questions

1 Read the following passage, and then answer the questions which follow.

Recycling waste

We all throw away a lot of rubbish. Bottles, cans, packets and gift wrappings all end up in the bin. An average North American family's annual waste weighs as much as the Statue of Liberty.

Most of this rubbish is tipped, sometimes to fill holes in the ground. But an increasing amount of it is being recycled. Paper waste can be recycled to produce paper towels and newsprint. Glass can be broken, remelted, and reused. Metals can be melted and reused. Vegetable waste can be composted and used to improve soil. Now the first plastic recycling schemes are being developed.

Britain uses 40000 tonnes of polyethylene terephthalate, or PET, each year. Most of it is used to make bottles for soft drinks. This plastic is not biodegradable. Plastic bottles thrown onto rubbish dumps stay there virtually for ever.

PET is recyclable. But profitable recycling plants need at least 4000 tonnes of PET each year. Other problems include difficulty in transporting the bulky bottles, and in separating the PET from other plastics with which it may be mixed. Nevertheless, PET is now beginning to be recycled in both the USA and Britain.

a Give examples of paper, glass, metal, vegetable and plastic waste.

b Which is the oldest type of recycling mentioned in the passage above?

c Give two uses of recycled paper.

d What is meant by 'biodegradable'?

e Biodegradable plastics are being developed. Although these would solve some of the problems of plastic disposal, they might create other difficulties. Can you suggest what these difficulties might be?

f What is the maximum number of profitable PET recycling plants which Britain could support?

g Why are there more problems in transporting waste PET bottles than in transporting waste glass bottles?

h What raw material resource is likely to be conserved by recycling PET? As well as being fuels, coal, oil and natural gas may be used as chemical resources.

a Name one chemical resource obtainable from each fossil fuel, and state how it is obtained from the fuel.

b Which fossil fuel has least use as a source of chemicals? Why is this?

c Certain chemicals can be obtained from both coal and oil. Explain why they are currently obtained from oil. Will this always be the case?

3 A wax candle is lit and an inverted empty jam jar is placed over it. The following observations are made.

- The candle soon goes out.
- Water vapour condenses on the inside of the jar.
- The air in the jar is found to contain more carbon dioxide than normal air after the candle goes out.

a Why does the candle go out?

b Candle wax is a compound. Which element in that compound produces water as it burns? Which element produces carbon dioxide as it burns?

c 'Candle wax is a hydrocarbon.' Explain.

d Name one other gas, which is normally only found in air as a pollutant, which will be found in the jar after the candle has gone out. Explain its presence.

4 a Draw structural formulae for:
  i ethane
  ii butane
  iii hexane.

b Which of these three compounds does not have isomers?

c Draw structural formulae for:
  i ethene
  ii propene.

d Explain why ethene undergoes addition polymerisation, while ethane does not.

5 Explain why the polymers listed below are used for the purposes stated.

a PTFE – artificial joint surfaces

b polystyrene – drinking cups

c PVC – electrical insulation

d polythene – washing-up liquid bottles

6 a Give details of experiments which demonstrate that magnesium is more reactive than copper.

b How is this reactivity relevant to the extraction of magnesium and copper from their ores?

c How could you establish how reactive iron is, relative to magnesium and copper?

d Copper ores are becoming scarce. What products will be affected as copper prices rise?

7 Our heavy reliance on iron and its alloys makes rusting a serious industrial and domestic problem.

a What measures can be taken to control the rusting of iron?

b Why is 'rusting' not a problem with aluminium?

c Why is aluminium a more expensive metal than iron, even though it is more plentiful?

d Apart from cost, name another advantage of iron and its alloys over aluminium.

e Name an advantage of aluminium and its alloys over iron.

8 Open-cast mining is a very efficient way of getting coal out of the ground. Britain has considerable reserves of coal which are shallow enough to be extracted by open-cast mining. Before open-cast mining can take place, planning permission has to be given. During a one year period from 1986 to 1987, planning applications were approved for open-cast mining over 4492 hectares of land, thought to hold 35658000 tonnes of coal. In the same year, applications for 2206 hectares were refused, amounting to 14220000 tonnes of coal. The total output of coal from open-cast mines was 14596000 tonnes during that time. Total reserves for which planning permission had been granted, were 77575000 tonnes.

a Draw up a table, chart or graph to display these figures clearly.

b What objections might people have to open-cast mining in:
  i urban areas
  ii agricultural areas
  iii national parks?

c At the end of 1987, how many years' reserves of coal were available for open-cast mining, assuming it continues at the present rate?

THE PERIODIC TABLE

This image of atoms was taken using a scanning tunnelling electron microscope.

The Periodic Table arranges elements in order of atomic number

In Topic 8, you saw how the Periodic Table lays out the hundred or so known elements in order of their atomic number. The atomic number of an element is the number of **protons** in the nucleus of one of its atoms. So the oxygen atom in Figure 79.1 has an atomic number of eight. It is the eighth element in the Periodic Table. Hydrogen has one proton in the nucleus of each of its atoms, so it is the first element in the table. Lawrencium has 103 protons, so it is the 103rd element.

But why is the Periodic Table such an odd shape? It has two elements in the first row or **Period,** eight in both the second and third, 18 in the fourth and 32 in the sixth. This sixth row is so long that elements 58 to 71 are normally taken out of it, so that the table fits conveniently onto a sheet of paper. Why are there not just ten elements in each row, giving a neat, square layout?

The answer is that the layout we use does not just contain the symbols of the 103 elements. It also contains a lot of information about them. Understanding the Periodic Table helps us to understand why the elements behave as they do.

Vertical groups contain elements with the same number of outer electrons

Atoms have tiny, negatively charged particles, **electrons**, which move in orbits around the nucleus. The number of electrons in an atom is the same as the number of protons. So a hydrogen atom has one electron, moving in an orbit around its nucleus. A helium atom has two electrons, both of which move in the same orbit. Lithium, the third element, has three electrons, two of which move in the first orbit near the nucleus. But the third electron moves in a second orbit, further out from the nucleus.

This is why lithium is in Period 2 of the Periodic Table. It has electrons in both its first and second orbits. Hydrogen

and helium are in Period 1 because they only have electrons in the first orbit.

All the elements in Period 2, like lithium, have electrons in the first and second orbits. The tenth element, neon, has ten electrons. It has two in the first orbit, and eight in the second. These electrons fill these two orbits. So the 11th element, sodium, with 11 electrons, has two electrons in the first orbit, eight in the second and one in the third. Because it has an electron in the third orbit, sodium begins Period 3 of the Periodic Table.

This means that every element with one electron in its outer orbit is at the start of a Period. It is in the first column, or first **Group** of the Periodic Table. Every element in the final Group has a completely full outer orbit. Every other Group consists of elements with the same number of outer electrons. Since the outer electrons determine how an element will behave, elements in the same Group behave similarly.

Questions

1 Copy and complete this table. The first line is done for you.

Element	Atomic number	Period	Group	Number of orbits occupied	Number of outer electrons
potassium	19	4	1	4	1
sulphur					
lead					
barium					
xenon					
bromine					

2 a What are the elements in Group 7?
  b Would you expect them to behave in a similar way? Explain your answer.
  c What are the elements in Period 3?
  d Would you expect them to behave in a similar way? Explain your answer.
3 a Draw diagrams to show the electron configurations of:
     i magnesium
     ii nitrogen
     iii phosphorus
     iv argon
     v calcium
     vi neon.
  b Which of these elements would you expect to behave similarly? (You should be able to find three pairs.)

Fig. 79.1 An oxygen atom

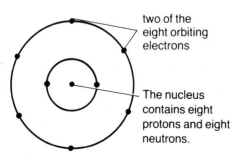

two of the eight orbiting electrons

The nucleus contains eight protons and eight neutrons.

Fig. 79.2 A lithium atom

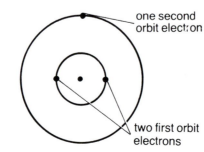

one second orbit electron

two first orbit electrons

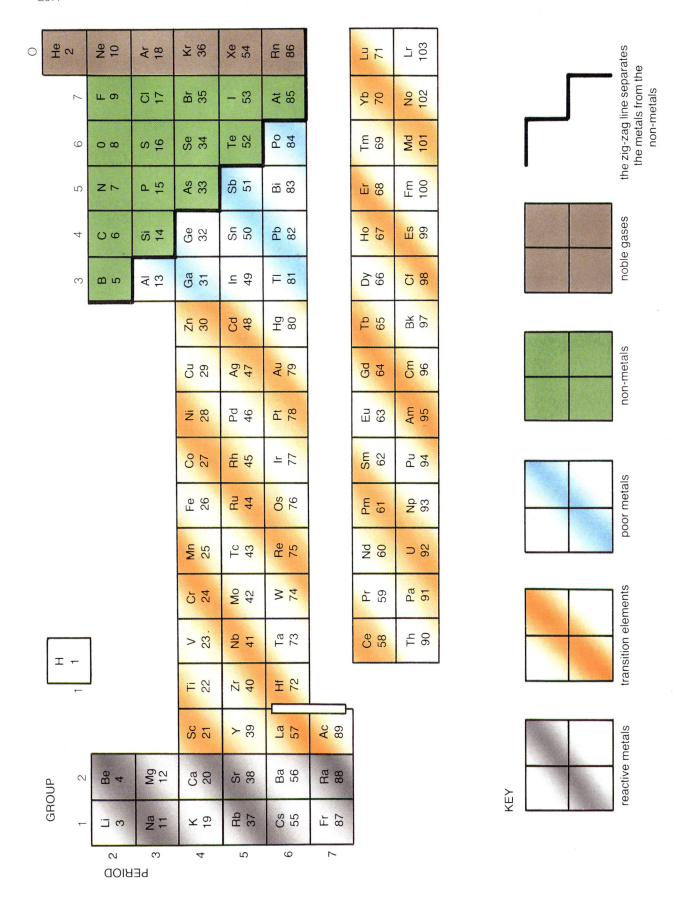

Fig. 79.3 The Periodic Table. The full names of the elements are listed on page 207.

80 THE ALKALI METALS

The elements in Group 1 of the Periodic Table behave similarly because they all have one outer electron.

Group 1 - the alkali metals

The six metals in Group 1 are **lithium**, **sodium**, **potassium**, **rubidium**, **caesium** and **francium**. The first three are commonly kept in schools, and your teacher may be able to demonstrate that they have the following properties:

- they all have low densities. Lithium, sodium and potassium will float on water.
- they are all soft and can be cut with a knife. Potassium is the softest of the three, and lithium is the hardest.
- they all have silvery, shiny surfaces when clean or freshly cut.
- they are all good conductors of heat and electricity.
- they all burn in air to form a solid oxide. These oxides all dissolve readily in water to form an alkaline solution of the metal hydroxide.
- they all react vigorously with water to give hydrogen gas and an alkaline solution of the metal hydroxide.

$$2Li(s) + 2H_2O(l) \rightarrow 2LiOH(aq) + H_2(g)$$
$$2Na(s) + 2H_2O(l) \rightarrow 2NaOH(aq) + H_2(g)$$
$$2K(s) + 2H_2O(l) \rightarrow 2KOH(aq) + H_2(g)$$

However, they do not all react with water with equal vigour. Lithium does not get hot enough to melt, while sodium and potassium do. Of these, sodium does not usually get hot enough to ignite the hydrogen formed, while potassium always does.

These properties, some of them very unusual for metals, are also shown by rubidium, caesium and francium. The alkali metals are truly a family of similarly behaving metals.

However, they are not identical. Some properties change gradually as you go down the Group. For example, all these metals are soft. But the softness increases from the top to the bottom of the Group. All react vigorously with water. But the vigour of the reaction increases as you go down the Group. Such gradual changes are called **trends**. They are useful because they enable us to make predictions about elements we have not actually observed in action. Rubidium, for example, is rarely seen in school laboratories, but you ought to be able to predict that it is softer than potassium and reacts more vigorously with water than potassium, as well as displaying all the other properties in the list above.

Fig. 80.1 Lithium reacting in cold water. The lithium is reacting steadily, but is not molten, and no flame has been produced.

Fig. 80.3 Potassium reacting in cold water. It is reacting very vigorously, and is molten. It has ignited the hydrogen being produced, causing the pink coloured flame.

Fig. 80.2 Sodium reacting in cold water. It is reacting rapidly and is so hot that it is molten. However, it has not ignited the hydrogen which is produced.

These metals behave similarly because they all have one outer electron

The Group 1 metals have very similar properties to each other, as you have read on the opposite page. Why is this? The answer lies in the arrangement of electrons in their atoms. They all have different numbers of electrons (lithium has 4, sodium has 11, potassium has 19) but if you write down their electron configurations you should notice a similarity. Lithium's electron configuration is 2, 1, sodium's is 2, 8, 1 and potassium's is 2, 8, 8, 1. They all have **one** electron in the outer orbit, so all these elements behave in a very similar manner. When these elements react, their atoms always lose one electron to form an ion with a 1+ charge, for example:

$$Na \longrightarrow Na^+ + e^-$$

Losing the single outer electron produces a sodium ion with an electron configuration of 2, 8. This ion has a complete outer orbit. All Group 1 elements have atoms that lose their single outer electrons easily, so they are all reactive.

Alkali metal compounds are similar to each other

You have already learned some of these similarities. For example:
• alkali metal oxides dissolve in water
• alkali metal hydroxides are alkalis.
But there are many more similarities between the compounds of these metals:
• alkali metal compounds are stable and do not decompose easily
• alkali metal compounds are ionic, so they have high melting and boiling points, they are soluble in water, they conduct electric current when they are molten or dissolved in water, and they form crystals
• most simple alkali metal compounds are white when solid and form colourless solutions in water.

We are very familiar with many alkali metal compounds from our daily lives. Some sodium compounds are particularly familiar.
• 'Bicarb', or to give it its proper name **sodium hydrogencarbonate**, is used as a raising agent in cookery (it is one of the ingredients in baking power) and is found in indigestion tablets.
• Washing soda consists of **sodium carbonate** crystals.
• Common salt, or **sodium chloride**, is the best known of all. It has been used for centuries in cookery to improve the flavour of food. In larger amounts salt will preserve food and stop it going bad. Adding salt lowers the freezing point of water, so we spread salt on our roads in winter to help stop them icing up.

Trends can be explained by considering atomic radius

We have said that all Group 1 metals are very reactive, but that doesn't mean they are all equally reactive. Remember there is a **trend** of reactivity with water; they get more reactive going down the Group. This must be because the single outer electron is more easily lost from the atoms of the elements further down the Group. The further down the Group we go, the more orbits of electrons each atom has. This makes the radius of each atom bigger; for example, a potassium atom has a bigger radius than a sodium atom. This means the outer electron is further from the nucleus, which means in turn it is more easily lost. So potassium is more reactive than sodium.

Fig 80.4 Salted cod.

Questions

1 a List as many properties of caesium as you can, taking trends in softness and reactivity with water into account.
b When an alkali metal reacts with water, why is the resulting solution alkaline?
c Which positive ion does each alkali metal form? Why is this?
d Estimate the melting point of potassium from the following data:

Element	lithium	sodium	rubidium	caesium
Melting point (°C)	180	98	39	29

2 The following data refer to the solubilities, in g per 100 g of water, of certain compounds of Group 2 metals.

Mg(OH)$_2$	0.0009	MgSO$_4$	33
Ca(OH)$_2$	0.156	CaSO$_4$	0.21
Sr(OH)$_2$	0.80	SrSO$_4$	0.013
Ba(OH)$_2$	3.9	BaSO$_4$	0.00024

a What trends are apparent here?
b Estimate the solubilities of
i Be(OH)$_2$
ii BeSO$_4$.
3 To answer this question, you will need to use general information to make predictions about a compound you may never have come across.
Would you expect potassium bromide to
a dissolve in water?
b have a low melting point?
c be covalently bonded?
d be easy to decompose?
e be green?
f conduct electricity when melted?
g be crystalline?

4 Find out about francium. Why did its discovery come a long time after the other elements in Group 1?

EXTENSION

Group 7 – the halogens

Group 7 consists of **fluorine**, **chlorine**, **bromine**, **iodine** and the extremely rare radioactive element **astatine**. These five elements, the halogens, have many things in common, as Table 81.1 shows (opposite).

- They are, without exception, non-metals.
- They form diatomic molecules, for example Cl_2 and Br_2.
- They form a simple ion with a 1⁻ charge, for example Cl^- and Br^-.
- They form acidic compounds with hydrogen, for example HCl and HBr.

Their other properties, while similar, show clear trends.

- All have low melting points and boiling points when compared with other elements. Their melting points and boiling points increase steadily down the Group.
- All are reactive elements. Their reactivity increases up the Group.
- All are coloured elements. Their colour increases down the Group.

You can see for yourself the trend in reactivity in the halogens if you try Investigation 81.1. You will not, however, be able to look at all of the halogens. Fluorine is a very reactive element, not used in school laboratories. It reacts explosively or in flames with many things, including hydrogen, alkanes and most metals. In similar circumstances, chlorine reacts much more slowly and less spectacularly, while bromine and particularly iodine are even slower to react. They may need considerable heating in order to react at all.

Halogen compounds

The halogens form many compounds. They form simple **salts** called **halides**. In fact the word 'halogen' means 'salt-former'. They form simple salts with the alkali metals (like sodium chloride, NaCl, sodium bromide, NaBr and sodium iodide, NaI). All these sodium salts are white crystalline solids that are soluble in water. Most of the salts formed by the halogens are soluble in water. However, this is not the case with their silver salts. If a solution of sodium chloride is added to a solution of silver nitrate, an **insoluble precipitate** of silver chloride forms. A similar reaction takes place if silver nitrate solution is added to solutions of sodium bromide or sodium iodide. The equations for these reactions are:

$$NaCl + AgNO_3 \longrightarrow AgCl + NaNO_3$$
$$NaBr + AgNO_3 \longrightarrow AgBr + NaNO_3$$
$$NaI + AgNO_3 \longrightarrow AgI + NaNO_3$$

Silver chloride, silver bromide and silver iodide are *not* soluble in water, which is why they form precipitates. When light hits a grain of one of these compounds it causes a grain of silver metal to form. This **light sensitivity** has led to the use of silver halides in making photographic film and photographic paper.

The halogens are similar because of their electron configurations

The reason for the similarity between the halogens is their electron configurations. All their atoms have seven electrons in their outer orbit. For example, fluorine's electron configuration is 2, 7. Chlorine's is 2, 8, 7. This means that the halogens have a strong tendency to gain one electron when they react. For example:

$$Cl + e^- \longrightarrow Cl^-$$

The addition of the the single outer electron gives a chloride ion with an electron configuration of 2, 8, 8. This ion has a complete outer orbit. All halogen atoms tend to gain one electron when they react, forming ions with a single negative charge. Electrons are gained more readily the nearer the outer orbit is to the nucleus. Therefore fluorine gains electrons most readily and is the most reactive of the halogens.

Fig. 81.1 Precipitates of, from the left, silver chloride, silver bromide and silver iodide. The photograph was taken before the precipitates had time to settle. The colours of the precipitates are described as white, buff and yellow respectively.

Table 81.1   The halogens

Name	Metal or nonmetal	Formula of element	Simple ion formed	Melting point (°C)	Boiling point (°C)	Reactivity	Colour	Hydrogen compound
fluorine	nm	F_2	F^-	−220	−188	exceptionally high	very pale yellow	HF – acidic
chlorine	nm	Cl_2	Cl^-	−101	−35	high	green-yellow	HCl – acidic
bromine	nm	Br_2	Br^-	−7	59	average	red-brown	HBr – acidic
iodine	nm	I_2	I^-	114	184	low	purple-black	HI – acidic

Fig. 81.2   Three of the halogens. The gas jar on the left contains pale green chlorine gas, while the one on the right contains red-brown bromine vapour and a little bromine liquid. The watch glass contains nearly-black iodine.

INVESTIGATION 81.1

Halogens reacting

1 Take six clean test tubes, and number them one to six.
2 Put 1 cm³ of chlorine water into tubes one and two.
  Put 1 cm³ of bromine water into tubes three and four.
  Put 1 cm³ of iodine, dissolved in potassium iodide solution, into tubes five and six.
3 Put 1 cm³ of potassium iodide solution into tubes one and three.
  Put 1 cm³ of potassium bromide solution into tubes two and five.
  Put 1 cm³ of potassium chloride solution into tubes four and six.
4 Record all your observations in a results table.

Questions

1 In which three tubes does no reaction take place? (Hint – no colour change.)
2 In the other three tubes, identify the coloured product, and deduce the second product.
3 Write balanced chemical equations for the reactions which take place.
4 What do these reactions tell us about the reactivity of chlorine, bromine and iodine? Explain your answer.
5 For each reaction, state:
  a what is being oxidised.
  b what is being reduced.
  Give reasons for your answers.

— *EXTENSION* —

Iodine is an antiseptic

Iodine is not very reactive, so it is relatively harmless to human beings, but it is reactive enough to react with and kill bacteria. It is therefore a useful antiseptic. Iodine solution is painted on wounds to help kill bacteria that may have got into them. If you have a medicine cabinet at home it may have a bottle of 'tincture of iodine' in it for this purpose.

Questions

1 Explain the connection between one or more of the halogens and
a the photographic industry
b tap water
c antiseptics
d toothpaste
e bleach
f pesticides
2 a What is the name by which the Group 7 elements are known?
b What are the elements in Group 7?
c Give one characteristic property of a Group 7 element.
d Give one trend shown by the Group 7 elements.

3 The element below iodine in the Periodic Table is astatine, At. Predict this element's properties as fully as you can.
4 Why do the Group 7 elements form an ion with a single negative charge?
5 a What is unusual about the silver halides?
b What trend of colour do the silver halides have?
c What colour would you expect silver fluoride to be?
d What colour would you expect silver astatide to be?

82 THE ARCHITECT OF THE PERIODIC TABLE

The Periodic Table was first put together by Mendeleev in 1869.

Mendeleev predicted the existence of undiscovered elements

The similar behaviour of elements in the same Group was a key factor in the first attempts at drawing up the Periodic Table. Many attempts were made to set out the known elements into an organised pattern. The present arrangement was suggested by the Russian scientist Dmitri Ivanovich Mendeleev in 1869. It was a work of great genius.

Mendeleev knew nothing of atomic structure, so he could not use atomic numbers or electron configurations to arrange the elements into a pattern. He did, however, know about the relative atomic masses of the elements. Just over fifty elements were known at that time. He arranged these elements in order of relative atomic mass, with the lightest first. A version of his table is shown in Figure 82.1.

Mendeleev's Table of the Elements

H																
Li	Be											B	C	N	O	F
Na	Mg											Al	Si	P	S	Cl
K	Ca	*	Ti	V	Cr	Mn	Fe	Co	Ni	Cu	Zn	*	*	As	Se	Br
Rb	Sr	Y	Zr	Nb	Mo	*	Ru	Rh	Pd	Ag	Cd	In	Sn	Sb	Te	I
Cs	Ba	*	*	Ta	W	*	Os	Ir	Pt	Au	Hg	Tl	Pb	Bi		

Fig. 82.1

As well as using atomic mass to put the elements in order, Mendeleev started a new row at such places as to put similar elements into groups according to the properties he knew they had. Sometimes, in order to make his table work, he boldly predicted that there must be other elements, not yet discovered. For example, after zinc, Zn, the next heaviest known element was arsenic. But arsenic did not fit into the same group as aluminium, because it wasn't like aluminium at all. It was much more like phosphorus. So Mendeleev put arsenic below phosphorus, and stated that two undiscovered elements lay between zinc and arsenic.

Mendeleev used the principles of similarities and group trends to predict how these two undiscovered elements would behave. They were discovered quite soon afterwards, gallium in 1875 and germanium in 1886. They behaved just as Mendeleev had suggested. These and other, then undiscovered, elements are represented in Mendeleev's table as asterisks. All have since been discovered, and fit the table perfectly.

Mendeleev put the elements in order of their relative atomic masses (A_r). He did, however, make one exception. He put tellurium ($A_r = 128$) before iodine ($A_r = 127$). This was a bold and far sighted decision. Knowing nothing of atomic number or electron configuration, Mendeleev abandoned the principle of A_r order so that he could keep similar elements in the same Group. Tellurium is, in many ways, similar to the elements above it, especially sulphur and selenium. Iodine fits in perfectly with the other members of Group 7 – the **halogens**.

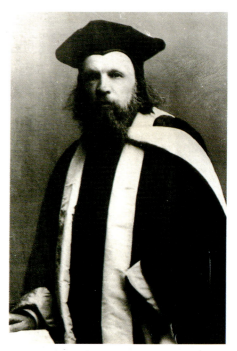

Fig. 82.2 The Russian chemist Dmitri Ivanovich Mendeleev. He was born in Siberia in 1834, and died in 1907. He was professor of chemistry at the University of St Petersburg.

DID YOU KNOW?

Caesium is used for accurate clocks. Caesium 133 is radioactive, producing 9 192 631 770 cycles of radiation in 1 s (plus or minus 20 cycles). The accuracy of a caesium clock is 8 parts in 10^{14}.

Group 0 – the rare gases

One complete group of elements is missing from Mendeleev's table. They were unknown to Mendeleev in 1869. These are the rare gases **helium**, **neon**, **argon**, **krypton**, **xenon** and **radon**. They make a Group on the far right of the Periodic Table.

These rare gases are another excellent example of a Periodic Table Group. They all have extremely similar chemistry. They are all highly unreactive, because they all have complete outer orbits of electrons. No compounds of helium, neon or argon have ever been formed. A few krypton compounds and a few xenon compounds are now known. So reactivity, such as there is, increases down this Group. There is also a trend of increased boiling point down the Group.

Since these gases are chemically unreactive, what uses they have are as elements. They are excellent for filling gas discharge tubes, where the different colours of light have made them popular for making advertising signs.

Element	Symbol	Boiling point (°C)
helium	He	−269
neon	Ne	−246
argon	Ar	−186
krypton	Kr	−152
xenon	Xe	−108

Table 82.1   Boiling points of the rare gases

Fig. 82.3   These fluorescent tubes contain argon.

Fig. 82.4   Some examples of transition metal compounds. From the top, looking from left to right, they are: copper (II) nitrate, cobalt (III) chloride, sodium dichromate (VI), manganese (II) chloride, iron (III) nitrate and nickel (II) chloride.

The transition metals

As you look across Period 2 from lithium to neon, or across Period 3 from sodium to argon, the properties of the elements are continually changing. In Periods 4, 5, 6 and 7, however, we find blocks of ten or more elements around the middle of each Period that are all very similar. These are elements 21–30 in Period 4, 39–48 in Period 5, 57–80 in Period 6 and 89–103 in Period 7. The elements in these blocks are called the **transition metals**. They have the properties listed below.

- They show the typical metal properties of high electrical conductivity, high levels of strength and high density. Most have high melting points.
- They form a wide range of brightly coloured compounds. The colour of these compounds is due to the transition metal ions they contain.
- Many have catalytic activity, for example:

$$2H_2O_2 \xrightarrow{MnO_2} 2H_2O + O_2$$

The catalyst is manganese (IV) oxide.

$$2SO_2 + O_2 \xrightarrow{V_2O_5} 2SO_3$$

The catalyst is vanadium (V) oxide.

$$N_2 + 3H_2 \xrightarrow{Fe} 2NH_3$$

The catalyst is iron.

- Most transition elements form more than one simple ion. For example, iron forms Fe^{2+} and Fe^{3+}.

Questions

1 a Why did Mendeleev leave two gaps after zinc?

b Was he correct to do so?

c The discovery of another element between, say, carbon and nitrogen would throw out the whole Periodic Table. Explain why such a discovery is not expected.

2 a What is the trend in reactivity of the rare gases?

b Scientists are trying to form compounds of the elements helium, neon and argon. With which of these are they most likely to be successful?

c Comment on the likely boiling point and reactivity of radon.

d The rare gases are also called the 'inert' or 'noble' gases. What do these words mean in this context?

3 Why did Mendeleev place tellurium before iodine in his table of the elements?

4 Titanium is element 22. Give five properties you would expect it to have.

83 THE PERIOD THREE ELEMENTS

The elements in any one Period in the Periodic Table do not behave in a similar way.

Elements in any one Period show a wide range of properties

The elements in each *Group* show many similarities. This is particularly true of the Groups at the extreme left and right of the Periodic Table – Groups 1, 2, 7 and 0. In a Group towards the middle of the Table, such as Group 4, we see more drastic changes as we descend the Group. In Group 4, we go from non-metals at the top (carbon and silicon) to metals at the bottom (germanium, tin and lead). There are many differences between carbon and lead, although there are some similarities too.

If we look at the elements in any one *Period*, however, we see very great differences as we cross the Period, and virtually no similarities between the first element and the last. If we take Period 3 as an example, it contains the elements sodium, magnesium, aluminium, silicon, phosphorus, sulphur, chlorine and argon. Going across the Period, the elements change from strongly metallic to strongly non-metallic. Their oxides change from being ionically bonded to being covalently bonded. The acidic or basic nature of their oxides also varies.

Table 83.1   Properties of the Period 3 elements

	$^{23}_{11}Na$	$^{24}_{12}Mg$	$^{27}_{13}Al$	$^{28}_{14}Si$	$^{31}_{15}P$	$^{32}_{16}S$	$^{35}_{17}Cl$	$^{40}_{18}Ar$
Melting Point (°C)	98	650	660	1410	44	113	–101	–189
Structure	metallic	metallic	metallic	giant covalent	simple molecular	simple molecular	simple molecular	single atoms
Formula of oxide	Na_2O	MgO	Al_2O_3	SiO_2	P_2O_5	SO_3	Cl_2O_7	none
Structure of oxide	ionic	ionic	ionic	giant covalent	simple molecular	simple molecular	simple molecular	—
Simple ion formed	Na^+	Mg^{2+}	Al^{3+}	none	P^{3-}	S^{2-}	Cl^-	none

---

INVESTIGATION 83.1

Investigating Period 3 oxides

In this investigation, you will be given solutions of Na_2O, P_2O_5 and SO_3, and solid samples of MgO, Al_2O_3 and SiO_2. Cl_2O_7 will not be investigated, because it is unpleasant and unstable.

WEAR SAFETY SPECTACLES

1 Take six clean test tubes. Label the test tubes one to six.
   Put 1 cm³ of Na_2O solution into tube one.
   Put 1 cm³ of distilled water and a little MgO powder into tube two.
   Put 1 cm³ of distilled water and a little Al_2O_3 powder into tube three.
   Put 1 cm³ of distilled water and a little SiO_2 powder into tube four.

Put 1 cm³ of P_2O_5 solution into tube five.
   Put 1 cm³ of SO_3 solution into tube six.
   Shake tubes two, three and four.
2 Add two drops of universal indicator solution to each tube. Record the pH of each solution.
3 If any of the tubes are at pH 7, this could be because the oxide is neutral, or the oxide could be an insoluble acid or an insoluble base. To find out, take a very little of the oxide and:
i see if it dissolves in a little hot sodium hydroxide solution. TAKE GREAT CARE! If it dissolves, then it

is acidic.
ii see if it dissolves in a little hot, dilute sulphuric acid. TAKE GREAT CARE! If it dissolves, then it is basic.

Questions

1 What happens to the pH of the oxides as you cross Period 3?
2 Complete the following sentence: 'In general, metal oxides are .... while non-metal oxides are ....'
3 SiO_2 will dissolve if boiled in very strong sodium hydroxide solution. What does this tell us about SiO_2?

The structures of matter

This expression means the ways in which the atoms that substances consist of are arranged. Nature has come up with several types of structure but only the four most important are explained on this page. Note as you read how important it is to link the structure to the properties of the substances – to connect in our minds the atoms we cannot see and the behaviour we *can* see and measure.

1 Metallic. The structure of around 80% of the elements and of the infinite number of mixtures (called **alloys**) that can be made from them is metallic. In a **metallic** structure the metal atoms are arranged in a regular repeating pattern or **lattice**. The metal atoms have all effectively lost their outer electrons, these electrons form a 'sea' of electrons that is part of the structure and able to move freely. The sea of mobile electrons binds the metal particles together strongly giving most metals strength and high melting points. It also makes metals good conductors of heat and electricity.

2 Simple molecular. A substance whose atoms are bonded together in small groups (or small **molecules)** by covalent bonds has a **simple molecular** structure. The covalent bonds are strong so that the atoms within the molecules are held together effectively. But the forces that attract molecules to their neighbours are weak, so these substances are either gases, liquids or solids that melt easily. They have no free electrons or other free electrically charged particles so they are bad conductors of electricity and heat. Note that the rare gases are a special example of this type of structure, consisting of individual atoms not bonded to any others, moving freely in a gaseous state.

3 Giant covalent. A substance whose atoms are all bonded into one vast array (or **lattice**) by covalent bonds has a **giant covalent** structure. All forces between particles are therefore strong covalent bonds and these substances are hard solids with high melting points. They have no free electrons or other free electrically charged particles so they are bad conductors of electricity and heat, although graphite is a partial exception to this being a reasonable conductor of electricity.

4 Ionic. Compounds with an ionic structure consist of two types of particle, one that has gained one or more electrons to become a **negative ion** and one that has lost one or more electrons to become a **positive ion**. The two different types of ion form a regular repeating pattern in which they alternate, called a lattice. Ionic compounds are therefore crystalline. The ions are held in place by the strong forces of attraction between each ion and its neighbours of opposite charge. Compounds with an ionic structure therefore have high boiling points. Since the ions are electrically charged, they will conduct electric current if they are free to move. This happens when the ionic compound is molten or dissolved in water.

Questions

1 Where in Period 3 do we find:
a non-metals?
b elements forming negative ions?
c elements with high melting points?
d elements forming positive ions?
e metals?
f elements with low boiling points?
g elements with acidic oxides?
h elements with basic oxides?
2 What is the structure of each of the following substances?
a magnesium oxide
b silicon
c sulphur
d sodium
e graphite
f sulphur trioxide
g silicon dioxide
h magnesium
i argon
j sodium chloride
k oxygen
l xenon
m aluminium oxide
n brass
o diamond
p gold

3 The table below contains some information about carbon and lead.

	Electrical conductivity	Acidic/basic nature of oxide	Bonding in chloride
Carbon	Graphite is a semiconductor. Diamond is a non-conductor.	Carbon dioxide is acidic, i.e. it can neutralise alkalis.	CCl_4 is covalent.
Lead	Good conductor	PbO_2 and PbO are amphoteric, i.e. they can neutralise acids and alkalis.	$PbCl_4$ is covalent, $PbCl_2$ is ionic.

How does this information show how the Group 4 elements change from non-metals to metals as the Group is descended?

Relative atomic masses enable us to work out relative molecular masses

The chemical industry needs to know about reacting amounts and about the amount of product being formed. Take, for example, the Thermit reaction:

$$2Al(s) + Fe_2O_3(s) \rightarrow Al_2O_3(s) + 2Fe(l)$$

How much aluminium and how much iron oxide have to be mixed so that there is exactly the right amount of each with no waste? How much iron would form? The answers start in the same place that Mendeleev started when laying out his periodic table — with the relative atomic masses of the elements.

Table 84.1 gives the relative atomic masses of some important elements. The table gives the comparative masses of individual atoms.

You can use these figures to work out the masses of molecules of elements or compounds.

For example:

An oxygen molecule, O_2, contains 2 oxygen atoms. The relative atomic mass of an oxygen atom is 16, so the relative molecular mass of oxygen is: 16 + 16 = 32.

A sulphur molecule, S_8, contains eight sulphur atoms. So its relative molecular mass is: 8 x 32 = 256.

A water molecule, H_2O, contains two hydrogen atoms and one oxygen atom. So its relative molecular mass is: 1 + 1 + 16 = 18.

A calcium carbonate molecule, $CaCO_3$, contains one calcium atom, one carbon atom and three oxygen atoms. So its relative molecular mass is: 40 + 12 + (3 x 16) = 100.

A magnesium nitrate molecule, $Mg(NO_3)_2$, contains one magnesium atom, two nitrogen atoms and six oxygen atoms. So its relative molecular mass is: 24 + (2 x 14) + (6 x = 148.

Relative molecular masses help us to work out mass equations

To find reactant and product amounts in actual reactions, we just need to do what we did on the left for every substance in the equation for the reaction. This gives us a 'mass equation'.

For example:

$$
\begin{array}{ccccccc}
Mg(s) & + & CuO(s) & \rightarrow & MgO(s) & + & Cu(s) \\
24 & + & (64+16) & \rightarrow & (24+16) & + & 64 \\
24 & + & 80 & \rightarrow & 40 & + & 64
\end{array}
$$

This mass equation has no units. If we put the same units onto *every* term in the mass equation – grams for example – it will still be true.

$$24\,g + 80\,g \rightarrow 40\,g + 64\,g$$

We now have something we can work with. When 24g of magnesium reacts with 80g of copper (II) oxide they produce 40g of magnesium oxide and 64g of copper.

24g Mg   80g CuO   40g MgO   64g Cu

Fig. 84.1

Total mass of reactants equals total mass of products

When writing mass equations, it is important to take into account any large numbers in the chemical equation. For example, in the following equation, the large number 2 in front of the formula for water means that two molecules of water react with each atom of calcium.

$$
\begin{array}{ccccccc}
Ca(s) & + & 2H_2O(l) & \rightarrow & Ca(OH)_2(aq) & + & H_2(g) \\
40 & + & 2 \times (1+1+16) & \rightarrow & 40+16+16+1+1 & + & 1+1 \\
40\,g & + & 36\,g & \rightarrow & 74\,g & + & 2\,g
\end{array}
$$

Now look back at the Thermit reaction. We can now answer the question that was asked at the beginning of this topic.

$$
\begin{array}{ccccccc}
2Al(s) & + & Fe_2O_3(s) & \rightarrow & Al_2O_3(s) & + & 2Fe(l) \\
27+27 & + & 56+56+16+16+16 & \rightarrow & 27+27+16+16+16 & + & 56+56 \\
54\,g & + & 160\,g & \rightarrow & 102\,g & + & 112\,g
\end{array}
$$

So 54g of aluminium reacts with 160g of iron (III) oxide to produce 102g of aluminium oxide and 112g of iron.

Notice that, in all these reactions, the total mass of the reactants equals the total mass of the products.

Atoms and molecules are counted in moles

An aluminium atom has a relative atomic mass of 27. An iron atom has a relative atomic mass of 56. **This means that 27g of aluminium and 56g of iron both contain the same number of atoms**. Since the relative atomic mass of carbon is 12, 27g of aluminium and 56g of iron both contain the same number of atoms as 12g of carbon.

What is this number? Since atoms are very small, the number is very large. In fact, it is approximately 600 000 000 000 000 000 000 000 or 6×10^{23}. The amount of substance containing this number of particles is called a **mole**.

The mass of a mole of atoms of an element is the element's relative atomic mass, in grams. For example:

Element	Relative atomic mass A_r	Mass of one mole of atoms
hydrogen	1	1g
calcium	40	40g
sulphur	32	32g

In the same way, the mass of a mole of molecules of an element or compound is the molecule's relative molecular mass in grams.

Substance	Formula	Relative molecular mass of one molecule	Mass of one mole in grams
hydrogen	H_2	$1 + 1 = 2$	2g
sulphur	S_8	$8 \times 32 = 256$	256g
water	H_2O	$1 + 1 + 16 = 18$	18g
carbon dioxide	CO_2	$12 + 16 + 16 = 44$	44g
ammonium sulphate	$(NH_4)_2SO_4$	$(2 \times 14)+(8 \times 1)+32+(4 \times 16)= 132$	132g

Notice that the mass of a mole of hydrogen atoms, H, is 1g, while the mass of a mole of hydrogen molecules, H_2, is 2g. Similarly, the mass of a mole of sulphur atoms, S, is 32g, while the mass of a mole of sulphur molecules, S_8, is 256g. It is important to say exactly which you are talking about!

Table 84.1 Relative atomic masses of some common elements

Element	Symbol	A_r
hydrogen	H	1
lithium	Li	7
carbon	C	12
nitrogen	N	14
oxygen	O	16
fluorine	F	19
sodium	Na	23
magnesium	Mg	24
aluminium	Al	27
phosphorus	P	31
sulphur	S	32
chlorine	Cl	35.5
potassium	K	39
calcium	Ca	40
iron	Fe	56
copper	Cu	64
zinc	Zn	65
bromine	Br	80
silver	Ag	108
iodine	I	127
lead	Pb	207

Fig. 84.2 Each watch glass contains one mole of substance. Starting at the top right, and working clockwise, they are:
one mole of copper sulphate molecules, $CuSO_4\ 5H_2O = 250\,g$
one mole of sulphur atoms, $S = 32\,g$
one mole of carbon atoms, $C = 12\,g$
one mole of sodium chloride molecules, $NaCl = 58.5\,g$
one mole of iron atoms, $Fe = 56\,g$
one mole of glucose molecules, $C_6H_{12}O_6 = 180\,g$.

Questions

1 a What is the relative atomic mass of one atom of each of these elements?
   i Na **ii** Mg **iii** Zn **iv** Cu **v** Pb **vi** I **vii** P
b What is the mass, in grams, of one mole of atoms of each of the above elements?
c What is the relative molecular mass of one molecule of each of the following substances?
   i I_2 **ii** Cl_2 **iii** P_4 **iv** KCl **v** $ZnBr_2$ **vi** H_3PO_4
   vii $Mg(OH)_2$ **viii** $Al(NO_3)_3$
d What is the mass, in grams, of one mole of molecules of each of the above substances?

2 Write mass equations for the following equations:
a $H_2SO_4(aq) + Mg(s) \rightarrow MgSO_4(aq) + H_2(g)$
b $2HCl(aq) + CuO(s) \rightarrow CuCl_2(aq) + H_2O(l)$
c $2HNO_3(aq) + CaCO_3(s) \rightarrow Ca(NO_3)_2 + H_2O(l) + CO_2(g)$
d $AgNO_3(aq) + NaCl(aq) \rightarrow AgCl(s) + NaNO_3(aq)$
e $Pb(NO_3)_2(aq) + H_2SO_4(aq) \rightarrow PbSO_4(s) + 2HNO_3(aq)$
f $2NaCl(l) \rightarrow 2Na(l) + Cl_2(g)$
g $2Al_2O_3(l) \rightarrow 4Al(l) + 3O_2(g)$
h $Fe_2O_3(s) + 3CO(g) \rightarrow 3CO_2(g) + 2Fe(l)$
i $2AgNO_3(aq) + Cu(s) \rightarrow Cu(NO_3)_2(aq) + 2Ag(s)$
j $C_{10}H_{22}(l) \rightarrow C_8H_{18}(l) + C_2H_4(g)$

85 REACTING AMOUNTS 2

If the amount of one reactant is changed, the amounts of other reactants and products will change too.

Half the reactant gives half the product

In Topic 84, you saw how relative atomic masses can be used to work out how much reactant and product is involved in a reaction. For example:

$$\begin{array}{ccccccc} Mg(s) & + & H_2O(g) & \rightarrow & MgO(s) & + & H_2(g) \\ 24 & + & 1+1+16 & \rightarrow & 24+16 & + & 1+1 \\ 24\,g & + & 18\,g & \rightarrow & 40\,g & + & 2\,g \end{array}$$

So, if you were asked 'How much water reacts with 24 g of magnesium?' the answer is easy. It is 18 g. If you were asked 'How much magnesium and water would you need to make 40 g of magnesium oxide?', the answer would be 24 g of magnesium and 18 g of water.

But what if you were asked how much water reacts with 12 g of magnesium, or 240 g, or 4 g? You can work out the answers like this:

> 12 g of magnesium is half of 24 g
> So you need half of 18 g = 9 g of water

> 240 g of magnesium is ten times 24 g
> So you need 10 x 18 g = 180 g of water

> 4 g of magnesium is $^1/_6$ of 24 g
> So you need $^1/_6$ of 18 g = 3 g of water.

You can also work out how much magnesium oxide would form in each case. For example:

> 12 g of magnesium is half of 24
> So you get half of 40 g = 20 g of magnesium oxide.

Try calculating the amounts of magnesium oxide formed if you used 240 g or 4 g of magnesium. You should find you get 400 g and 6.67 g of magnesium oxide respectively.

1 Magnesium reacting with steam, which is coming [fro]m wet mineral wool plug on the right. Both products – [hydroge]n which is burning and white magnesium oxide smoke [are visi]ble on the left.

A worked example

Sulphur trioxide is made by combining sulphur dioxide with oxygen. The equation is:

$$2SO_2(g) + O_2(g) \rightarrow 2SO_3(g)$$

1 How much oxygen combines with the following masses of sulphur dioxide:
  a 128 g   **b** 32 kg?
2 How much sulphur trioxide would form in each case?
  To answer these questions, we first need a mass equation:

$$\begin{array}{ccccc} 2SO_2(g) & + & O_2(g) & \rightarrow & 2SO_3(g) \\ 2 \times (32+16+16) & + & 16+16 & \rightarrow & 2 \times (32+16+16+16) \\ 128\,g & + & 32\,g & \rightarrow & 160\,g \end{array}$$

1 a The mass equation tells us the answer to this directly: 32 g of oxygen combine with 128 g of sulphur dioxide.
  b So long as the units are the same throughout, it doesn't matter which units of mass are used in a mass equation. We can work in kilograms throughout the calculation.
  32 kg of sulphur dioxide is $^1/_4$ of 128 kg ($^{32}/_{128} = {}^1/_4$).
  So $^1/_4$ of 32 kg of oxygen is needed = **8 kg of oxygen**.
2 a The mass equation tells us directly: 160 g of sulphur trioxide form from 128 g of sulphur dioxide.
  b 32 kg of sulphur dioxide is $^1/_4$ of 128 kg.
  So $^1/_4$ of 160 kg of sulphur trioxide will form
  = **40 kg of sulphur trioxide**.

Gas volumes can be calculated

Where gases are involved, we can calculate the volume of the gas, too. This is because one mole of molecules of any gas occupies 24 dm³ at normal room temperature and normal atmospheric pressure. (A volume of 24 dm³ is 24 000 cm³ or 24 l – about a suitcase full.) Table 85.1 states this fact for five gases, but it is true for *all* gases.

So, for any reaction with gaseous reactants or products, we can now write the *volumes* (of the gaseous substances only) in with the mass equation. For example:

$$\begin{array}{ccccccccc} CaCO_3(s) & + & 2HCl(aq) & \rightarrow & CaCl_2(aq) & + & H_2O(l) & + & CO_2(g) \\ (40+12+ & + & 2(1+35.5) & \rightarrow & (35.5+ & & (1+1 & + & (12+16 \\ 16+16+16) & & & & 35.5+40) & & +16) & & +16) \\ 100\,g & + & 73\,g & \rightarrow & 111\,g & + & 18\,g & + & 44\,g \text{ or } 24\text{ dm}^3 \end{array}$$

Or:

$$\begin{array}{ccccc} N_2(g) & + & 3H_2(g) & \rightarrow & 2NH_3(g) \\ (14+14) & + & 3(1+1) & \rightarrow & 2(14+1+1+1) \\ 28\,g & + & 6\,g & \rightarrow & 34\,g \\ 24\,\text{dm}^3 & + & (3 \times 24)\,\text{dm}^3 & \rightarrow & (2 \times 24)\,\text{dm}^3 \end{array}$$

Notice here that *three* moles of hydrogen react, so they occupy 3 x 24 = 72 dm³. Two moles of ammonia occupy 2 x 24 = 48 dm³. Note also that the total volume of gaseous products *need not* equal the total volume of gasous reactants.

EXTENSION

A worked example Table 85.1

What volume of carbon dioxide is produced when 1.24g of copper carbonate react with excess hydrochloric acid?

(Using excess hydrochloric acid ensures that all of the 1.24g of $CuCO_3$ will react.)

Start with the mass equation:

Gas	Mass of one mole	Volume of one mole
CO_2	12 + 16 + 16 = 44g	24dm³
H_2	1 + 1 = 2g	24dm³
O_2	16 + 16 = 32g	24dm³
N_2	14 + 14 = 28g	24dm³
NH_3	14 + 1 + 1 + 1 = 17g	24dm³

$$CuCO_3(s) + 2HCl(aq) \rightarrow CuCl_2 + H_2O(l) + CO_2(g)$$
$$124g + 73g \rightarrow 135g + 18g + 44g \text{ or } 24dm^3$$

From the mass equation above, you can see that 124g of $CuCO_3$ produces 24dm³ of CO_2.

So 1.24g of $CuCO_3$, which is 1/100 of 124, will produce 24/100 dm³ which is 0.24dm³ or **240 cm³ of CO_2**.

Fig. 85.2 Copper carbonate reacting with hydrochloric acid which is going blue as copper chloride forms.

Working out the percentage of nitrogen in a fertiliser

You don't need to be an industrial chemist to want to know the answers to this kind of question. Some of these ideas are also of use to people such as farmers. If a soil is low in natural nitrogen compounds, then it becomes less fertile. A farmer may need to add a nitrogen-containing fertiliser to the soil. If the farmer had the choice of potassium nitrate, ammonium sulphate or ammonium nitrate, and wanted to use the one with the highest percentage of nitrogen, which should be chosen? The answer to this question can be worked out like this:

For all three substances: % nitrogen = $\dfrac{\text{mass of nitrogen in one molecule}}{\text{mass of one molecule}} \times 100$

For potassium nitrate, KNO_3, this is: $\dfrac{14}{39 + 14 + 16 + 16 + 16} \times 100 = 13.9\%$

For ammonium sulphate, $(NH_4)_2SO_4$, it is: $\dfrac{14 + 14}{(2 \times 14) + (8 \times 1) + 32 + (4 \times 16)} \times 100 = 21.2\%$

For ammonium nitrate, NH_4NO_3, it is: $\dfrac{14 + 14}{14 + (4 \times 1) + 14 + (3 \times 16)} \times 100 = 35.0\%$

So ammonium nitrate contains the highest percentage of nitrogen. But the farmer might choose instead to use a fertiliser like potassium nitrate, if potassium also needed to be added to the soil.

Fig. 85.3 Fertiliser often contains sources of nitrogen, phosphorus and potassium – N, P and K. Fertilisers are manufactured with different ratios of these three important elements, for use with different soils or different crops.

Questions

1 For the reaction:
$TiCl_4(l) + 2Mg(s) \rightarrow Ti(s) + 2MgCl_2$
the mass equation is:
190g + 48g → 48g + 190g.

a What is the relative atomic mass of titanium (Ti)?

b What is the mass of one mole of titanium?

c What mass of titanium chloride reacts with 48g of magnesium?

d What mass of titanium forms in the process?

e What mass of magnesium reacts with the following amounts of titanium chloride:
i 19g **ii** 380g **iii** 190kg?

f What mass of titanium forms in each of the reactions in part e?

2 Sodium hydrogen carbonate reacts with dilute sulphuric acid, according to this equation:
$2NaHCO_3(s) + H_2SO_4(aq) \rightarrow$
$Na_2SO_4(aq) + 2CO_2(g) + 2H_2O(l)$

a Write a mass equation for this reaction.

b What mass of sulphuric acid reacts with the following amounts of sodium hydrogen carbonate:
i 56g **ii** 14g **iii** 0.336g?

c What mass of sodium sulphate forms in each case in part b?

d What mass of carbon dioxide forms in each case in part b?

e What volume of carbon dioxide forms in each case in part b?

f How much sodium hydrogen carbonate must be put into excess sulphuric acid in order to produce the following amounts of carbon dioxide:
i 12dm³
ii 8dm³
iii 100cm³?

3 The following are all copper compounds:
i CuO **iv** CuS
ii $CuCO_3$ **v** Cu_2O
iii $Cu(OH)_2$

a Find the percentage of copper in each compound.

b What is the maximum amount of copper obtainable from 100 tonnes of each compound?

FINDING FORMULAE

The formula of a compound can be calculated from experimental results.

We need to know the amount of each element in a compound

By now you know the formulae of a lot of chemical compounds. The mass equations and the calculations based on them in Topics 84 and 85 would have been impossible without knowing the formulae of the substances involved. All these formulae are based on experimental results. Before deciding what the formula of a particular substance is scientists found, by investigation, how much of each of its constituent elements it contains.

Some worked examples

We can find the formula of magnesium oxide by the following method.

Accurately weigh out 24 g of magnesium ribbon and put it in a crucible with a lid on. Heat strongly with a Bunsen burner so that all the magnesium burns and forms magnesium oxide. Weigh the amount of magnesium oxide in the crucible when all the magnesium has reacted. It weighs 40 g.

Word equation magnesium + oxygen $\longrightarrow$ magnesium oxide
From results   24 g   +   ?g $\longrightarrow$ 40 g

All of the gain in mass from 24 g to 40 g must have been due to the oxygen that the magnesium combined with. Therefore the amount of oxygen involved must have been 40 g − 24 g = 16 g.

Now we know
24 g magnesium + 16 g oxygen $\longrightarrow$ 40 g magnesium oxide

To find the formula of the magnesium oxide we must find out how many atoms of magnesium react with how many atoms of oxygen. To do this we convert the mass of each reactant into moles.

A_r magnesium = 24
so one mole of magnesium atoms weighs 24 g
A_r oxygen = 16 so one mole of oxygen atoms weighs 16 g

So one mole of magnesium atoms reacts with one mole of oxygen atoms.

Therefore one magnesium atom reacts with one oxygen atom.

Therefore the formula of magnesium oxide is MgO.

Using a similar method it is found that when lithium reacts with oxygen to form lithium oxide 1.4 g of lithium will burn to give 3.0 g of lithium oxide.

Therefore
1.4 g lithium + 1.6 g oxygen $\longrightarrow$ 3.0 g lithium oxide
A_r lithium = 7, A_r oxygen = 16

To convert a reacting mass into a number of moles we divide the reacting mass by the mass of one mole of the substance.

1.4 g of lithium is 0.2 moles of lithium (1.4 g ÷ 7 g = 0.2)

1.6g of oxygen is 0.1 moles of oxygen (1.6 g ÷ 16 g = 0.1)

0.2 moles of lithium reacts with 0.1 moles of oxygen. The ratio of these numbers is 2 : 1. (A ratio must be in whole numbers.)

Therefore 2 moles of lithium reacts with 1 mole of oxygen.
Therefore 2 atoms of lithium reacts with 1 atom of oxygen.
Therefore the formula of lithium oxide is Li_2O.

1.08 g of aluminium can form 2.04 g of aluminium oxide. What is the formula of aluminium oxide?

1.08 g aluminium + 0.96 g oxygen
$\longrightarrow$ 2.04 g aluminium oxide

A_r aluminium = 27, A_r oxygen = 16

1.08 g of aluminium is 0.04 moles of aluminium
(1.08 g ÷ 27 g = 0.04)

0.96g of oxygen is 0.06 moles of oxygen
(0.96 g ÷ 16 g = 0.06)

0.04 moles of aluminium reacts with 0.06 moles of oxygen. The ratio of these numbers is 2 : 3. (A ratio must be in whole numbers.)

Therefore 2 moles of aluminium reacts with 3 moles of oxygen.

Therefore 2 atoms of aluminium reacts with 3 atoms of oxygen.

Therefore the formula of aluminium oxide is Al_2O_3.

Questions

1   18.2 g of an oxide of vanadium contains 10.2 g of vanadium.
   a How many grams of oxygen does it contain?
   b A_r vanadium = 51, A_r oxygen = 16
      How many moles of vanadium have combined with how many moles of oxygen?
   c How many atoms of vanadium have combined with how many atoms of oxygen?
   d What is the formula of the vanadium oxide?

2   Sodium and oxygen can react together in various proportions depending on the conditions. Three different oxides can be made with three different formulae.
   i 15.5 g of sodium oxide contains 11.5 g of sodium
   ii 7.8 g of sodium peroxide contains 4.6 g of sodium
   iii 16.5 g of sodium superoxide contains 6.9 g of sodium.
   a Calculate the formulae of sodium oxide sodium peroxide, and sodium superoxide.
   b Only one of your answers should fit in with what you know about sodium forming an ion with a 1+ charge and oxygen forming an ion with a 2− charge. Find out what the explanation is for the other two formulae.

EXTENSION

Another experimental example

This next example starts from the amount of compound and then works out the amount of each element it contains.
Suppose you take a little porcelain boat of mass 101.22 g. You put some copper oxide powder into it, so that its mass is now 102.02 g. You heat it in a stream of hydrogen, which takes all the oxygen away so that the copper oxide is converted to copper. When you reweigh the boat containing the copper, you find its mass is now 101.86 g. Was the copper oxide, copper (I) oxide, Cu_2O, or copper (II) oxide, CuO?

The mass of the powder used was 102.02 − 101.22 = 0.80 g. This powder contained 102.22 − 101.86 = 0.16 g of oxygen. The powder contained 0.80 − 0.16 = 0.64 g of copper.

So 0.64 g of copper were combined with 0.16 g of oxygen in the original powder.
1 mole of copper atoms has a mass of 64 g.

So 0.64 g of copper is: $\frac{0.64}{64}$ = 0.01 moles of atoms.

1 mole of oxygen atoms has a mass of 16 g.

So 0.16 g of oxygen is: $\frac{0.16}{16}$ = 0.01 moles of atoms

So 0.01 moles of copper atoms were combined with 0.01 moles of oxygen atoms.

So 1 mole of copper atoms was combined with 1 mole of oxygen atoms.

So 1 atom of copper was combined with 1 atom of oxygen.

So the formula of the copper oxide was **CuO**.

Formulae can be calculated from percentage data.
Sometimes when a compound is analysed the various amounts of the elements it is made of are expressed as percentages. The formula of the compound can be calculated from this data.

Example: Calcium fluoride is 51.3% calcium and 48.7% fluorine by mass. If A_r calcium = 40 and A_r fluorine = 19 calculate the formula of calcium fluoride.

i divide each percentage by the A_r of that element

51.3 ÷ 40 = 1.28 calcium   48.7 ÷ 19 = 2.56 fluorine

ii calculate the ratio of calcium to fluorine

1.28 : 2.56 is 1 : 2 so the formula of calcium fluoride is CaF_2.

Example: A hydrocarbon used as camping gas is 81.82% carbon and 18.18% hydrogen by mass. If A_r carbon is 12 and A_r hydrogen is 1 calculate the formula of the hydrocarbon.

i divide each percentage by the A_r of that element
81.82 ÷ 12 = 6.82 carbon   18.18 ÷ 1 = 18.18 hydrogen

ii calculate the ratio of carbon to hydrogen. It isn't obvious! When a ratio isn't obvious divide both numbers by the smaller one:

6.82 ÷ 6.82 = 1 carbon   18.18 ÷ 6.82 = 2.67 hydrogen

You should recognise this as the whole number ratio 3 : 8. The formula of the hydrocarbon is C_3H_8 and its name is propane.

a 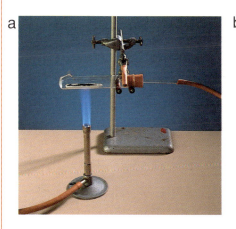 b 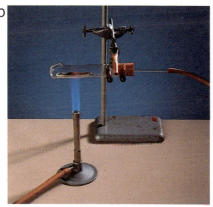 c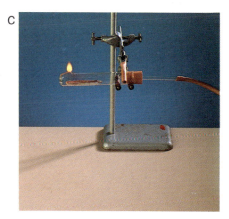

Fig. 86.1   Copper oxide in a porcelain boat is heated in a stream of hydrogen gas. Excess hydrogen can be seen burning at the hole in the boiling tube. After a little while only copper remains. Hydrogen is kept flowing while the copper cools, to prevent hot copper from re-oxidising on contact with the air.

Questions

1   Some lead oxide was put into a massed porcelain boat, and the mass of the boat plus the oxide was recorded. The oxide was then heated in a stream of hydrogen, until only lead was left. The final mass was then recorded. The results are given below. (Relative atomic mass of lead = 207; relative atomic mass of oxygen = 16.)
   Mass of empty boat = 131.00 g
   Mass of boat + lead oxide = 135.78 g
   Mass of boat + lead = 135.14 g.
   a What mass of lead oxide did the boat contain?
   b What mass of lead did the lead oxide contain?

c How many moles of lead is this?
d What mass of oxygen did the lead oxide contain?
e How many moles of oxygen is this?
f Deduce the formula of the lead oxide.
2   A compound is 40% calcium, 12% carbon and 48% oxygen by mass. If A_r calcium is 40, A_r carbon is 12, and A_r oxygen is 16 calculate its formula.
3   A compound is 20.2% aluminium and 79.8% chlorine by mass. If A_r aluminium is 27 and A_r chlorine is 35.5 calculate its formula

Copper can be purified by electrolysis

So far you have come across electrolysis as a property of ionic compounds when molten or in solution, as a way of extracting sodium and chlorine from sodium chloride and as a way of extracting aluminium from molten aluminium oxide. In all of these examples the electrodes used were inert, which means that as far as possible the electrodes do not take any part in the reactions that take place on them. The electrodes are just there to conduct the electric current into and out of the electrolyte. But some substances do not make inert electrodes however, and if they are used as electrodes they do take part in the reactions. An example of this is if copper sulphate is electrolysed using copper electrodes.

At the cathode copper ions gain electrons and form copper metal on the electrode:

$$Cu^{2+} + 2e^- \longrightarrow Cu$$

At the anode copper atoms from the electrode lose electrons and form copper ions in solution:

$$Cu \longrightarrow Cu^{2+} + 2e^-$$

Since the second reaction is just the reverse of the first there seems to be little point in it. However the copper which plates on the cathode is highly pure – much purer than the copper that the anode was originally made of. Electrolysing copper sulphate solution using a pure copper cathode and a large impure copper anode is therefore used as a way of purifying copper. Any impurities contained in the copper anode do not go into the solution or plate the cathode. Instead they form a sludge at the bottom of the tank. Although this sludge consists of 'impurities' it isn't worthless. Two impurities that can be found in copper are silver and gold, so the sludge may be quite valuable.

The amount of copper purified can be calculated

The purification of copper must be able to work at a profit or it wouldn't be done. Electricity is expensive. The purification company must know that the amount of pure copper it will turn out (plus any profit from sludge) is worth the expense. How can the amount of copper purified be calculated?

It is clear from the half equation $Cu^{2+} + 2e^- \longrightarrow Cu$ that two electrons flowing round the circuit cause one copper atom to be deposited. Therefore it follows that two moles of electrons are needed to deposit one mole of copper atoms. One mole of electrons is contained in 96 500 coulombs of electricity. The 'coulomb' is the unit used for amount of electric charge. The number of coulombs that have flowed around a circuit can be calculated by multiplying the current in amperes by the time in seconds.

Worked example 1

A copper sulphate solution is electrolysed using a pure copper cathode and a lump of impure copper as the anode. A current of 2 amperes is used for a total of 2000 seconds. How much pure copper is deposited on the cathode?

i The amount of charge that flowed = current × time
$$2\,A \times 2000\,s = 4000 \text{ coulombs}$$

ii Two moles of electrons are needed to deposit one mole of copper. Two moles of electrons are contained in $2 \times 96\,500$ coulombs of charge.
One mole of copper weighs 64.5 grams.
64.5 g of copper are deposited by 193 000 coulombs of charge.

iii If this 193 000 coulombs is scaled down to 4000 coulombs we have the answer to our problem.
64.5 g of copper from 193000 coulombs.
$64.5 \div 193\,000$ g of copper from 1 coulomb.
$64.5 \div 193\,000 \times 4000$ g of copper from 4000 coulombs.
1.34 g of copper from 4000 coulombs.

So the answer is that 4000 coulombs will purify 1.34 grams of copper.

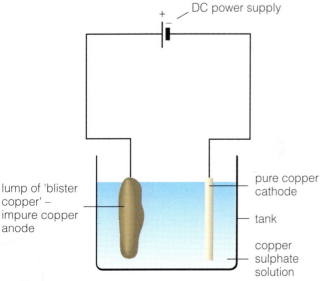

lump of 'blister copper' – impure copper anode

DC power supply

pure copper cathode

tank

copper sulphate solution

Fig. 87.1 Copper purification by electrolysis. The lump of impure copper ('blister copper' as it is called) gets smaller as pure copper builds up on the cathode. The solution is labelled copper sulphate solution but the important thing is that it contains Cu^{2+} ions.

A more complicated example

A copper sulphate solution is electrolysed using a pure copper cathode and a lump of impure copper as the anode. A current of 5 amperes is used. How long must the process be left to run if we want 1 kg of copper to be purified?

i   64.5 g of copper are deposited by 193 000 coulombs of charge.

ii 1 g of copper is deposited by $193\,000 \div 64.5$ coulombs of charge.

iii 1 kg of copper is deposited by $193\,000 \div 64.5 \times 1000$ coulombs of charge.
   i.e. 1 kg of copper is deposited by 3 000 000 coulombs of charge.

iv Charge = current × time
   3 000 000 coulombs = 5A × time
   $3\,000\,000 \div 5$ = time
   600 000 secs = time

So 600 000 secs would be needed to deposit 1 kg of copper if the current was 5A. This is nearly a week!

We can calculate the amount of substance that forms at one electrode if we know the amount that forms at the other electrode.

For other electrolysis reactions, where inert electrodes are used, a different product forms at each electrode. For example when molten sodium chloride is electrolysed the equation is:

$$2NaCl \longrightarrow 2Na + Cl_2$$

Given the fact that the A_r of sodium is 23 and the A_r of chlorine is 35.5 we can deduce that $2 \times (23 + 35.5)$g of sodium chloride gives 2×23 g of sodium and 2×35.5 g of chlorine on being electrolysed or:

$$117\,g\ NaCl \longrightarrow 46\,g\ Na + 71\,g\ Cl_2$$

So if we are told that during a period of production 46 kg of sodium are produced we can say that 71 kg of chlorine are produced at the same time. We can also work out what volume of gas this is. We know that one mole of gas occupies 24 dm³ at normal temperature and pressure. Since one mole of chlorine gas weighs 71 g the amount we have here is 1000 moles. This amount of gas would occupy 24 000 dm³.

Worked example 2

What volume of chlorine forms at the anode if copper chloride solution is electrolysed with inert electrodes and 6.45 g of copper is plated on the cathode? (A_r Cu = 64.5 A_r Cl = 35.5)

i   equation $CuCl_2 \longrightarrow Cu + Cl_2$

ii so $135.5\,g\ CuCl_2 \longrightarrow 64.5\,g\ Cu + 71\,g\ Cl_2$

iii Since 64.5 g of copper forms at the same time as 71 g of chlorine, 6.45 g of copper forms at the same time as 7.1 g of chlorine. This is 1/10 of a mole of gas and occupies a volume of 2.4 dm³ at normal temperature and pressure.
   answer = 2.4 dm³

Questions

1 a Silver is purified by a method very similar to the method used to purify copper.
   i   What solution do you think is used when silver is purified?
   ii What substance is the cathode made of when silver is purified?
   iii What substance is the anode made of when silver is purified?
   iv Draw a simple labelled diagram to show how silver is purified.

 b A silver ion has a charge of 1+. Its symbol is Ag⁺. The A_r of silver is 108.
   i   Write a half equation to show what happens at the cathode during the purification of silver.
   ii Write a half equation to show what happens at the anode during the purification of silver.
   iii How much silver would be purified by the passage of a current of 10 amperes for 20 minutes?
   iv How long would it take to purify 10.8 g of silver if a current of 8 amperes was used?

2 a Write an equation for the electrolysis of molten sodium chloride.
 b Write a mass equation for this process. (A_r Na = 23, A_r Cl = 35.5)
 c Calculate the mass of chlorine gas that will be produced when the following masses of sodium are produced by the electrolysis of molten sodium chloride:
   i   2.3 g
   ii 46 g
   iii 115 g.
 d What volume of chlorine would form in each case?

3 a Write an equation for the electrolysis of molten aluminium oxide.
 b Write a mass equation for this process. (A_r Al = 27, A_r O = 16)
 c What mass of aluminium would form at the cathode when the following volumes of oxygen are formed at the anode at normal temperature and pressure? (1 mole of oxygen occupies 24 dm³ at NTP)
   i   36 dm³
   ii 7.2 dm³
   iii 6 dm³
   iv 19 dm³.

CONCENTRATION

The amount of a substance dissolved in 1 dm³ is called the concentration.

Concentration is number of moles divided by volume

Scientists often look at reactions that involve solutions of reacting substances. Using a solution of a reactant in water is advantageous in many ways.

- Solutions allow reactant particles to meet and collide much more easily than if undissolved solids are used. For example, there is virtually no reaction between solid sodium chloride and solid silver nitrate. In solution, however, there is an immediate reaction:

$$NaCl(aq) + AgNO_3(aq) \longrightarrow AgCl(s) + NaNO_3(aq)$$

- Solutions of gases are easier to handle than the gases themselves. For example, hydrogen chloride in solution is easier to handle than hydrogen chloride gas.
- Some dangerous chemicals are safer to handle when diluted. Pure sulphuric acid, for example, is a thick syrupy liquid, and very unpleasant. It is known as concentrated sulphuric acid. Dilute sulphuric acid, $H_2SO_4(aq)$, must still be treated with respect, but it is much less dangerous.
- Having made solutions of the substances to be used, exact amounts can be measured out by volume, which is often more convenient than weighing.

But how can we know how much of our measured amount is reactant and how much is water? The answer involves the **concentration** of the solution.

The concentration of a solution is:

the number of moles of solute contained in each decimetre cubed of solution.

A decimetre cubed is abbreviated to dm^3. 1 dm^3 is 1000 cm^3, or 1 l.

So the units of concentration are **moles per dm³** or **mol/dm³**. (This used to be abbreviated to M. You may still come across this abbreviation occasionally.)

So a solution of sodium chloride of concentration $1 mol/dm^3$ contains one mole of sodium chloride dissolved in each dm^3 of solution. There are, 23 + 35.5 = 58.5 g, of sodium chloride per dm^3 of solution.

Fig. 88.1
The volumetric flask contains exactly 500 cm³ of sodium dichromate (VI) solution. The amount of sodium dichromate (VI) it contains is 26.2 grams, the same as the amount on the watch glass. Given the data A_r Na = 23, A_r Cr = 52, and A_rO = 16, calculate the concentration of the solution.

Worked examples

To find the concentration of a solution, divide the number of moles dissolved, by the total volume of solution in dm^3.

1 117 g of sodium chloride, NaCl, are used to make 4 dm^3 of salt solution. Find the concentration.

1 mole of NaCl has mass of 58.5 g

So 117 g is: $\frac{117}{58.5}$ = 2 moles of NaCl

Concentration = $\frac{\text{number of moles}}{\text{volume}}$

which is $\frac{2 \text{ moles}}{4 \text{ dm}^3} = \frac{1}{2}$

Therefore the concentration of the solution = 0.5 mol/dm³.

2 10 g of sodium hydroxide, NaOH, are used to make 1 dm^3 of sodium hydroxide solution. Find the concentration.

1 mole of NaOH has mass of 23 + 16 + 1 = 40 g

So 10 g is: $\frac{10}{40}$ = 0.25 moles of NaOH

Concentration = $\frac{\text{number of moles}}{\text{volume}}$

which is $\frac{0.25 \text{ moles}}{1 \text{ dm}^3}$ = 0.25

Therefore the concentration of the solution = 0.25 mol/dm³.

3 40 g of ammonium nitrate, NH_4NO_3, are used to make 250 cm³ of ammonium nitrate solution. Find the concentration.

1 mole of NH_4NO_3 has mass of 14 + 4 + 14 + 48 = 80 g
So 40 g contain 0.5 moles of NH_4NO_3

$250 \text{ cm}^3 = \frac{250}{1000}$ = 0.25 dm^3

Concentration = $\frac{\text{number of moles}}{\text{volume}}$

which is $\frac{0.5 \text{ moles}}{0.25 \text{ dm}^3}$ = 2

Therefore the concentration of the solution = 2 mol/dm³.

4 A bottle of wine contains 750 cm³ of wine. The wine contains 73 g of alcohol. The formula of alcohol is C_2H_5OH. Find the concentration of alcohol in the wine.

1 mole of alcohol has mass of 24 + 5 + 16 + 1 = 46 g

So 73 g is $\frac{73}{46}$ = 1.6 moles of alcohol

$750 \text{ cm}^3 = \frac{750}{1000}$ = 0.75 dm^3

Concentration = $\frac{\text{number of moles}}{\text{volume}}$

which is $\frac{1.6 \text{ moles}}{0.75 \text{ dm}^3}$ = 2.1

Therefore the concentration of alcohol in the wine is 2.1 mol/dm³.

Number of moles is volume x concentration

If we know the concentration of a solution, we can also work out how many moles of dissolved solute there are in any particular volume.

Number of moles = concentration x volume

1 How many moles of H_2SO_4 are there in 2 dm^3 of a solution of concentration 4 mol/dm^3?
Number of moles = concentration x volume
$\qquad = 4 \times 2 = 8$ moles.

2 How many moles of sodium hydroxide are there in 50 cm^3 of a solution of sodium hydroxide of concentration 0.1 mol/dm^3?

50 cm^3 = 0.05 dm^3

Number of moles = concentration x volume
$\qquad = 0.1 \times 0.05 = 0.005$ moles.

3 How many grams of pure nitric acid HNO_3 are there in 300 cm^3 of dilute nitric acid of concentration 0.6 mol/dm^3?

300 cm^3 = 0.3 dm^3

Number of moles = concentration x volume
$\qquad = 0.6 \times 0.3 = 0.18$ moles

1 mole of HNO_3 = 1 + 14 + 48 = 63 g

0.18 moles HNO_3 = 63 $\times$ 0.18 = 11.34 g
Therefore the mass of pure nitric acid is 11.34 g.

Volume is number of moles ÷ concentration

If we have a solution of a known concentration and we know how many moles of solute we need in our sample, we can calculate the volume required by dividing the number of moles by the concentration.

1 You need 0.1 moles of HCl. You have a 2 mol/dm^3 solution of HCl(aq). What volume should you take?

$$\text{volume} = \frac{\text{moles}}{\text{concentration}} = \frac{0.1}{2} = 0.05 \text{ dm}^3$$

So you should take 0.05 dm^3 or 50 cm^3 of the acid.

2 You need 0.5 moles of NH_3. You have a 0.5 mol/dm^3 solution of NH_3(aq). What volume should you take?

$$\text{volume} = \frac{\text{moles}}{\text{concentration}} = \frac{0.5}{0.5} = 1 \text{ dm}^3$$

So you should take 1 dm^3 of the ammonia solution.

3 What volume of a 0.10 mol/dm^3 sodium chloride solution should you take if you want it to contain exactly 2.0 g of dissolved salt?

1 mole of NaCl = 23 + 35.5 = 58.5 g

So 2.0 g is $\dfrac{2.0}{58.5} = 0.034$ moles

$$\text{volume} = \frac{\text{moles}}{\text{concentration}} = \frac{0.034}{0.10} = 0.34 \text{ dm}^3$$

So you should take 0.34 dm^3 or 340 cm^3 of the salt solution.

Questions

1 How many moles of H_2SO_4 are there in:
a 2 dm^3 of a 1 mol/dm^3 solution of H_2SO_4?
b 3 dm^3 of a 0.2 mol/dm^3 solution of H_2SO_4?
c 50 cm^3 of a 2 mol/dm^3 solution of H_2SO_4?
d 150 cm^3 of a 0.1 mol/dm^3 solution of H_2SO_4?

2 What volume of 0.1 mol/dm^3 HCl is needed if you require the sample to contain:
a 1 mole of dissolved HCl?
b 0.01 moles of dissolved HCl?
c 0.2 moles of dissolved HCl?
d 0.0004 moles of dissolved HCl?

3 What is the concentration of each of the following solutions?
a 2 moles of $MgCl_2$ is used to make 6 dm^3 of solution.
b 0.01 moles of $AgNO_3$ is used to make 2 dm^3 of solution.
c 0.5 moles of $CuSO_4$ is used to make 4 dm^3 of solution.
d 20 g of $CuSO_4$ is used to make 100 cm^3 of solution.
e 1 g of NaOH is used to make 25 cm^3 of solution.

4 In an experiment, 20 cm^3 of 0.1 mol/dm^3 HCl(aq) are found to be just enough to react with 5 cm^3 of 0.4 mol/dm^3 NH_3(aq).
a How many moles of HCl reacted?
b How many moles of NH_3 reacted?
c How many moles of HCl react with 1 mole of NH_3?
d How many molecules of HCl react with 1 molecule of NH_3?
e Write an equation for this reaction.

5 How many grams of dissolved hydrochloric acid HCl are there in the following samples of dilute hydrochloric acid (A_r H = 1, A_r Cl = 35.5)
a 25 cm^3 of 0.2 mol/dm^3 HCl
b 300 cm^3 of 2.5 mol/dm^3 HCl
c 70 cm^3 of 0.25 mol/dm^3 HCl.

6 What volume of 0.3 mol/dm^3 ammonia, NH_3, solution contains
a 10.2 g of dissolved ammonia
b 5 g of dissolved ammonia
c 3.6 dm^3 of dissolved ammonia gas (measured at normal temperature and pressure)
d 9 dm^3 of dissolved ammonia gas (measured at normal temperature and pressure)
(A_r N = 14, A_r H = 1, one mole of gas occupies 24 dm^3 at NTP).

7 25 cm^3 of sodium hydroxide solution of concentration 0.2 mol/dm^3 was put into a conical flask and some indicator was added. Hydrochloric acid solution was added to the flask slowly until the indicator showed the sodium hydroxide had just been neutralised. 18.2 cm^3 of hydrochloric acid was necessary to do this.
a Write an equation for the reaction between sodium hydroxide solution and hydrochloric acid.
b Calculate the number of moles of sodium hydroxide in 25 cm^3 of 0.2 mol/dm^3 solution.
c Look at the equation you did for part (a) and use your answer to part (b) to say how many moles of dissolved hydrochloric acid there must have been in the 18.2 cm^3 that was necessary to neutralise the sodium hydroxide.
d Calculate the concentration of the hydrochloric acid solution.
e This practical technique is called titration. Find out what use titration is and what apparatus is used.

Questions

1 The equation for iron reacting with bromine is:

$$2Fe(s) + 3Br_2(l) \rightarrow 2FeBr_3(s).$$

a Write a mass equation for this reaction.
b What mass of iron reacts with the following masses of bromine:
   i 240 g **ii** 96 g **iii** 0.720 g?
c What mass of iron bromide forms in each case in part b?
d What masses of iron and bromine must react to form 5.92 g of iron bromide?

2 One similarity between the Group 2 metals is seen in the formulae of the sulphates:

   $BeSO_4$ $MgSO_4$ $CaSO_4$ $SrSO_4$ $BaSO_4$

a Why do Group 2 metals all form a sulphate with the general formula MSO_4?
b Calculate the relative molecular mass of one molecule of each compound.

3 Where in the Periodic Table would you expect to find elements with the following properties:
a metal, soft, very reactive, forms an ion with a 1^+ charge?
b gas, totally unreactive?
c metal, hard and strong, forms a 1^+ ion and a 2^+ ion?
d non-metal, reactive, forms an ion with a 1^- charge?
e non-metal, but with some metallic properties. Formula of chloride is XCl_4?
f metal, dense, forms many coloured compounds, is a catalyst for certain reactions?

EXTENSION

4 Electrolysing saturated sodium chloride solution in a diaphragm cell or flowing mercury cell produces sodium hydroxide solution, hydrogen gas and chlorine gas according to the following equation:
   $2NaCl(aq) + 2H_2O(l) \rightarrow 2NaOH(aq) + H_2(g) + Cl_2(g)$
a Write a mass equation for this reaction, including the volumes of hydrogen and chlorine which form.
b If 40 tonnes of sodium hydroxide are made by this process:
   i what mass of chlorine is produced?
   ii what volume of chlorine is produced?
   iii what mass of sodium chloride is consumed?
c If 20 kg of hydrogen is produced by this process:
   i what volume of hydrogen is produced?
   ii what volume of chlorine is produced at the same time?
   iii what mass of sodium hydroxide is produced?

5 A fertiliser is made by mixing equal masses of potassium nitrate, KNO_3, and ammonium phosphate, $(NH_4)_3PO_4$.
   What is the percentage by mass in the fertiliser of:
a phosphorus?
b nitrogen?
c potassium?

6 The equation for aerobic respiration is:
   $6O_2 + C_6H_{12}O_6 \rightarrow 6CO_2 + 6H_2O$
a If one mole of oxygen is used in this reaction, how many moles of carbon dioxide will be produced?
b If 10 cm³ of oxygen is used, what volume of carbon dioxide will be produced?

The diagram shows a piece of apparatus which can be used for comparing the respiration rates of organisms in different conditions.

If the animals take in gas, the coloured liquid in the U-tube moves up the left-hand arm. If they give out gas, the liquid moves up the right-hand arm. The difference in the liquid level between the two arms of the manometer is proportional to the volume of gas released or produced.

~ " the woodlice respire aerobically, what would you
    ›ect to happen to the height of the liquid in the U-
)e?

d The experiment is usually performed with potassium hydroxide solution at the bottom of the tube containing the woodlice. Potassium hydroxide solution absorbs carbon dioxide, so any carbon dioxide produced occupies no extra volume in the apparatus.
   If potassium hydroxide is used, and the woodlice continue to respire aerobically, what would you expect to happen to the level of the liquid in the U-tube?
e Describe how you could use this apparatus to investigate the effect of temperature on the rate of respiration of woodlice.
f Temperature affects the volume of a gas. Would this affect the results of the experiment you have described in part e? If you think it would, what could you do to try to eliminate this problem?
g Could this apparatus be used to investigate the rate of photosynthesis of a small plant? Explain your answer fully.

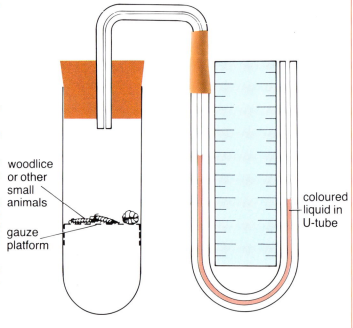

woodlice or other small animals

gauze platform

coloured liquid in U-tube

The Elements

Element	Symbol	Element	Symbol
Actinium	Ac	Mercury	Hg
Aluminium	Al	Molybdenum	Mo
Americium	Am	Neodymium	Nd
Antimony	Sb	Neon	Ne
Argon	Ar	Neptunium	Np
Arsenic	As	Nickel	Ni
Astatine	At	Niobium	Nb
Barium	Ba	Nitrogen	N
Berkelium	Bk	Nobelium	No
Beryllium	Be	Osmium	Os
Bismuth	Bi	Oxygen	O
Boron	B	Palladium	Pd
Bromine	Br	Phosphorus	P
Cadmium	Cd	Platinum	Pt
Caesium	Cs	Plutonium	Pu
Calcium	Ca	Polonium	Po
Californium	Cf	Potassium	K
Carbon	C	Praseodymium	Pr
Cerium	Ce	Promethium	Pm
Chlorine	Cl	Protactinium	Pa
Chromium	Cr	Radium	Ra
Cobalt	Co	Radon	Rn
Copper	Cu	Rhenium	Re
Curium	Cm	Rhodium	Rh
Dysprosium	Dy	Rubidium	Rb
Einsteinium	Es	Ruthenium	Ru
Erbium	Er	Samarium	Sm
Europium	Eu	Scandium	Sc
Fermium	Fm	Selenium	Se
Fluorine	F	Silicon	Si
Francium	Fr	Silver	Ag
Gadolinium	Gd	Sodium	Na
Gallium	Ga	Strontium	Sr
Germanium	Ge	Sulphur	S
Gold	Au	Tantalum	Ta
Hafnium	Hf	Technetium	Tc
Helium	He	Tellurium	Te
Holmium	Ho	Terbium	Tb
Hydrogen	H	Thallium	Tl
Indium	In	Thorium	Th
Iodine	I	Thulium	Tm
Iridium	Ir	Tin	Sn
Iron	Fe	Titanium	Ti
Krypton	Kr	Tungsten	W
Lanthanum	La	Uranium	U
Lawrencium	Lr	Vanadium	V
Lead	Pb	Xenon	Xe
Lithium	Li	Ytterbium	Yb
Lutetium	Lu	Yttrium	Y
Magnesium	Mg	Zinc	Zn
Manganese	Mn	Zirconium	Zr
Mendelevium	Md		

APPARATUS LISTS

Quantities given are those needed per group if the experiment is performed by pupils, or the quantity needed to perform a demonstration.

It is expected that teachers will provide safety goggles whenever pupils are handling potentially dangerous materials, or heating substances.

1.1 Using a smoke cell to see Brownian motion
smoke cell (including light source, lens, cell and cover slip)
power pack to suit bulb
leads
microscope; check that there is sufficient clearance between stage and objectives to accommodate smoke cell
waxed paper straw or string to make smoke

3.1 How quickly do scent particles move?
Pupils will ask for their own apparatus, but are likely to need:
perfume – any will do
stopcock
metre ruler or tape

3.2 Diffusion of two gases
The apparatus is shown in the diagram on page 13. Concentrated hydrochloric acid and ammonia solution give the best results, but care must be taken not to expose pupils to fumes at close quarters, or for very long. A small amount of each liquid can be poured into a watch glass in a fume cupboard, and a piece of cotton wool (held with forceps) dipped into each. If the soaked cotton wool is quickly pushed into the tube ends, and then immediately held in position with a bung, only a relatively mild smell will be noticed by pupils.

3.3 How small are potassium permanganate particles?
10 t   n rack
            of potassium perman-
tte to measure 1 cm^3

4.1 Cooling wax
Bunsen burner, mat tripod, gauze
Pyrex beaker
pieces of wax (paraffin)
test tubes and holders
retort stand, boss and clamp
access to water

4.2 Measuring the melting point of a solid
As for 4.1, but not wax
stearic acid
thermometer
This experiment could also be used to check the purity of different substances. Suitable examples might include $CaCl_2$ $6H_2O$ (m.p. 30 °C); hard paraffin wax (52 to 56 °C); butter (28 to 33 °C); lard (36 to 40 °C).

11.1 Comparing the strength of different metals
Samples of wire to include some or all of: copper, nichrome, iron, constantan or Eureka, tinned copper. The wires should be of at least 4 different thicknesses per metals.
masses and hangers (a hook of 24 swg copper can support about 5.5kg)
retort stands, clamps and bosses
something for masses to fall onto, e.g. a carpet tile

11.2 Breaking a wire
Samples of wire as 11.1
masses and hangers (a hook of 24 swg copper can support about 5.5kg)
retort stand, clamp and boss
hand lens
something for masses to fall onto, e.g. a carpet tile

17.1 Comparing ionic and covalent compounds
safety spectacles, spatula
A sodium chloride, ceramic paper
B copper sulphate, tongs
C zinc oxide, 1,1,1-trichloroethane
D sugar, test tubes in rack
E paraffin wax, 100 cm^3 beaker
F cetyl alcohol, circuit for testing for electrical conductivity, Bunsen burner

20.1 Measuring the percentage of oxygen in air
apparatus as in diagram
small pieces of copper wire, or copper turnings
two retort stands, clamps and bosses to support barrels of gas syringes

25.1 Finding the solubility of different substances
Students will ask for their own apparatus. They are likely to need:
test tubes and a rack
measuring cylinder, pipette or syringe for measuring volumes of water
stirring rod
distilled water
spatula
access to top pan balance
suitable substances to investigate include:
various sugars, such as glucose, sucrose, fructose etc; sodium chloride (0.3 g per gram of water), potassium sulphate (0.111 g per gram of water), copper carbonate (virtually insoluble)

26.1 How does detergent affect surface tension?
small piece of fabric such as cotton, cotton/polyester mix etc.
dropper pipette
access to water
small amount of liquid detergent such as washing-up liquid
This experiment can also be performed using a bird's feather. The effect of thoroughly washing the feather with a detergent could be investigated, and the implications for the treatment of oiled seabirds can then be considered.

28.1 Strong and weak acids
12 clean test tubes
2 short strips of magnesium ribbon
2 small marble chips
Universal indicator solution and colour chart (full range or pH 1–7 is best)
0.1 mol/dm^3 hydrochloric acid
0.1 mol/dm^3 ethanoic acid
10 cm^3 measuring cylinder

28.2 Measuring the pH of soil
At least two different samples of soil. The local soil should be used, as well as something which will contrast with it. Chalk soil, loam, peat and a commercially produced compost could be tried.
4 or 5 test tubes with corks to fit
small filter funnel and papers
rack or beaker to support tubes
syringe or measuring cylinder to measure 10 cm^3
access to water of pH approximately 7
spatula
Universal indicator paper or solution with colour chart

29.1 Neutralising a weak acid with a weak alkali
2 clean boiling tubes
1 clean teat pipette
full range universal indicator solution and colour chart
0.1 mol/dm^3 ethanoic acid
0.1 mol/dm^3 ammonia solution
glass rod

29.2 Neutralising a strong acid with a strong alkali
a clean burette
clean conical flask
thermometer
2 mol/dm^3 hydrochloric acid
2 mol/dm^3 sodium hydroxide solution
clean 20 cm^3 measuring cylinder or pipette
Bunsen burner

31.1 Metal hydroxide + acid →
metal salt + water
clean burette
clean 25 cm^3 pipette and filler
clean conical flask
0.1 mol/dm^3 sodium hydroxide solution
0.1 mol/dm^3 hydrochloric acid
phenolphthalein indicator

31.2 Ammonia solution + acid → ammonium salt + water
As for 31.1, but substitute ammonia solution for sodium hydroxide solution, and screened methyl orange for phenolphthalein.

31.3 Metal oxide + acid → metal salt + water
beaker
2 mol/dm^3 sulphuric acid
copper oxide
spatula
glass rod
Bunsen burner etc. for warming acid
thermometer
universal indicator paper
filter paper and funnel

31.4 Metal carbonate + acid → metal salt + water + carbon dioxide
beaker
2 mol/dm^3 sulphuric acid
magnesium carbonate powder
spatula
glass rod
filter paper and funnel

31.5 Metal + acid → metal salt + hydrogen
beaker
2 mol/dm^3 hydrochloric acid
magnesium ribbon or powder
spatula if powder used
glass rod

32.1 The effect of sulphur dioxide on plants
4 margarine tubs or other containers, with drainage holes
general purpose compost, e.g. John Innes Number 2
viable barley and maize seeds
means of labelling containers
4 watch glasses
cotton wool
sodium metabisulphite solution
access to water
4 large, transparent polythene bags
ties or rubber bands for bags

34.1 Exothermic and endothermic changes
beaker
25 cm^3 measuring cylinder, or pipette and filler
spatula
thermometer
stirring rod
copper sulphate solution
2 mol/dm^3 hydrochloric acid
iron filings
potassium sulphate
ammonium nitrate
magnesium ribbon
anhydrous copper sulphate
access to water

37.1 The thiosulphate cross
0.1 mol/dm^3 sodium thiosulphate
1 mol/dm^3 hydrochloric acid
distilled water
100 cm^3 beaker
50 cm^3 measuring cylinder
10 cm^3 measuring cylinder
paper with dark pencil cross drawn on it
stopclock

37.2 Factors affecting the rate of reaction
7 marble chips of similar size
measuring cylinder
apparatus as in diagram
thermometer
100 cm^3 beaker
pestle and mortar
500 cm^3 of 2 mol/dm^3 hydrochloric acid
stopclock
Bunsen burner, mat, tripod and gauze
crystallising dish

38.1 Catalysing the decomposition of hydrogen peroxide with an enzyme
20 volume hydrogen peroxide solution
syringe, measuring cylinder or pipette and filler to measure 20 cm^3
4 boiling tubes
glass marking pen
fresh liver
kitchen knife
tile
pestle and mortar
Bunsen burner, mat, tripod and gauze
glass beaker
stopclock
forceps
stirring rod
splint
Fresh liver reacts very fast, and it may be found that the froth rises too quickly to be measured. Most living tissues contain enough catalase to make this experiment work successfully, so pieces of apple or potato, or a yeast suspension, could be used as an alternative.

41.1 The reaction between magnesium and sulphuric acid
100 cm^3 measuring cylinder
50 cm^3 measuring cylinder
gas generator and delivery tube (see diagram)
crystallising dish
magnesium ribbon
access to balance
0.1 mol/dm^3 sulphuric acid
distilled water
stopclock

41.2 The digestion of starch by amylase
3 boiling tubes
glass marking pen
1% starch solution
0.1% amylase solution
10 cm^3 syringe
water bath at 35 °C
test tube rack
access to refrigerator
spotting tile
iodine in potassium iodide solution, with dropper
glass rod
stopclock
thermometer
Amylase from different sources behaves differently, and it is worth checking the best concentration beforehand.
If no refrigerator is available, students could use a beaker of iced water as a water bath for tube C.

47.1 Crystallisation
microscope and light source
salol powder
microscope slide
Bunsen burner

49.1 Rock hardness
small samples of a range of minerals of known hardness, plus other 'common equivalents' from the chart
one or more unknown minerals

52.1 The effect of acid on rocks
Small samples of various rocks, including some or all of the following: limestone, marble, slate, granite or basalt, pumice, sandstone, a piece of brick, a piece of concrete and your local stone.
glass Petri dish, or tile
dropper pipette
safety goggles
dilute hydrochloric acid

53.1 Measuring the percentage of water and humus in a soil sample
evaporating dish
pen to mark evaporating dish
soil sample, freshly taken
balance to mass soil samples
oven set to around 70–80 °C
method of heating soil samples to a much higher temperature – either a very hot oven, or Bunsen burner and rod for stirring

53.2 The effect of lime on clay
small sample of clay – potter's clay will do
calcium oxide powder (lime)
test tube and test tube rack
access to water

55.1 Erosion and deposition by waves
Students will ask for their own apparatus, but are likely to need:
a shallow tray, such as a large seed tray or laboratory storage tray
coarse sand, washed to remove small particles (if these are left in, they form a suspension, so it is difficult to see what is happening)
fine gravel
access to water
a piece of water-resistant card, wood, or a ruler to make waves
stop watch or clock
pieces of wood or plastic to make groynes
protractor
This investigation can be left very open-ended, or students can be limited to investigating the effects of just one variable. In either case, it is good to encourage the collection of some quantitative data.

61.1 Which elements are present in oil products?
apparatus shown in Figure 61.1
kerosine
hexane
ethanol
candles
lime water
means of lighting the kerosine and other oil products

63.1 Chemical properties of alkanes and alkenes
cyclohexane
cyclohexene
crucible lid
matches or Bunsen and tapers or some other means of lighting the chemicals
heatproof mat
two teat pipettes
four test tubes
bromine water
acidified potassium manganate (VII) solution
four corks

65.1 Properties of polymers
access to a fume cupboard
dilute hydrochloric acid
test tubes
marker pen or some other means of labelling test tubes
Bunsen burner
paper bag
polythene bag
wool
acrylic
samples of a range of other polymers, which should include polystyrene sheet (e.g. a drinking cup) and phenolic resin (e.g. a broken plug)
scissors
retort stand, boss and clamp
suitable masses
beakers
sodium chloride

68.1 The reaction of metals with acids
hydrochloric acid, 1 mol/dm^3
magnesium
iron
copper
lead
zinc – granules, shot, coarse filings or ribbon
test tubes

70.1 The reactivity of silver
copper sulphate solution, 0.1 mol/dm^3
silver nitrate solution, 0.1 mol/dm^3
wide gauge copper wire
silver wire or strip
emery cloth
two test tubes
means of marking test tubes

70.1 Do air and water affect the rate at which iron rusts?
5 dry boiling tubes
means of marking boiling tubes
5 iron nails
emery cloth
cotton wool
anhydrous calcium chloride or silica gel
two bungs to fit tubes
olive oil
safety glasses
access to gas supply and Bunsen burner
access to water

77.1 Comparing different methods for preventing corrosion

The following metal samples could be tested: steel pieces coated with oil, paint or tin, steel nails coated with brass or zinc. Students will need two samples each, with the coating on one damaged by scratching, hack-sawing, or cleaning.

iron nails

pieces of stainless steel

test tubes and rack

access to water

various acids

magnesium ribbon

tinned fruit

81.1 Halogens reacting

6 clean test tubes and a test tube rack

means of marking test tubes

2 cm^3 of each of the following:

chlorine water, made by bubbling chlorine gas into water in a fume cupboard

bromine water, made by dissolving 1 cm^3 liquid bromine in 1000 cm^3 water

iodine in potassium iodide solution – 1g solid iodide in 0.1 mol/dm^3 potassium iodide

potassium iodide solution – 0.1 mol/dm^3

potassium bromide solution – 0.1 mol/dm^3

potassium chloride solution – 0.1 mol/dm^3

pipettes or measuring cylinders for measuring 1 cm^3 of the various solutions

83.1 Halogens reacting

6 clean test tubes and a test tube rack

means of marking test tubes

6 clean boiling tubes

MgO, Al$_2$O$_3$ and SiO$_2$ powders

NaOH(aq) labelled 'NaO solution' – 2 mol/dm^3

H$_2$SO$_4$(aq) labelled 'SO$_2$ solution' – 1 mol/dm^3

universal indicator solution and colour chart

NaOH(aq) labelled normally – 2 mol/dm^3

H$_2$SO$_4$(aq) labelled normally – 1 mol/dm^3

distilled water prechecked at pH7

ANSWERS TO QUESTIONS

Topic	Questions		Topic	Questions		Topic	Questions	
5	**1**	10	85 (cont.)	**1fi**	4.8 g	88	**1a**	2
				1fii	96 g		**1b**	0.6
13	**page 38**	1 million chloride ions		**1fiii**	48 g		**1c**	0.1
				2bi	32.67 g		**1d**	0.015
	page 39	2 million chloride ions		**2bii**	8.17 g		**2a**	10 dm^3
				2biii	0.196		**2b**	100 cm^3
				2ci	47.33 g		**2c**	2 dm^3
34	**1a**	2640		**2cii**	11.83 g		**2d**	4 dm^3
	1b	3338		**2ciii**	0.284 g		**3a**	0.33 mol/dm^3
	1c	698		**2di**	29.33 g		**3b**	0.005 mol/dm^3
	2a	2328		**2dii**	7.33 g		**3c**	0.125 mol/dm^3
	2b	2252		**2diii**	0.176 g		**3d**	1.25 mol/dm^3
	2c	76		**2ei**	16 dm^3		**3e**	1 mol/dm^3
				2eii	4 dm^3		**4a**	0.002 moles
84	**1ai**	23		**2eiii**	0.096 dm^3 or 96 cm^3		**4b**	0.002 moles
	1aii	24		**2fi**	42 g		**4c**	1 mole
	1aiii	65		**2fii**	28 g		**4d**	1 molecule
	1aiv	64		**2fiii**	0.35 g		**5a**	0.1825 g
	1av	207		**3ai**	80%		**5b**	27.4 g
	1avi	127		**3aii**	51.61%		**5c**	0.639 g
	1avii	31		**3aiii**	65.31%		**6a**	2 dm^3
	1bi	23 g		**3aiv**	66.67%		**6b**	980 cm^3
	1bii	24 g		**3av**	88.89%		**6c**	500 cm^3
	1biii	65 g		**3bi**	80 tonnes		**6d**	1.25 dm^3
	1biv	64 g		**3bii**	51.61 tonnes		**7a**	0.005
	1bv	207 g		**3biii**	65.31 tonnes		**7b**	0.005
	1bvi	127 g		**3biv**	66.67 tonnes		**7c**	0.275 mol/dm^3
	1bvii	31 g		**3bv**	88.89 tonnes			
	1ci	254						
	1cii	71						
	1ciii	124	86 (page 200)	**1a**	8 g			
	1civ	74.5		**1b**	0.20, 0.5			
	1cv	225		**1c**	2, 5			
	1cvi	98						
	1cvii	58	(page 201)	**1a**	4.78 g			
	1cviii	151		**1b**	4.14 g			
	1di	254 g		**1c**	0.02			
	1dii	71 g		**1d**	0.64 g			
	1diii	124 g		**1e**	0.04			
	1div	74.5 g						
	1dv	225 g	87	**1biii**	13.43 g			
	1dvi	98 g		**1biv**	20 min 6 sec			
	1dvii	58 g		**2ci**	3.55 g			
	1div	151 g		**2cii**	71 g			
				2ciii	177.5 g			
85	**1a**	48		**2di**	1.2 dm^3			
	1b	48 g		**2dii**	24 dm^3			
	1c	190 g		**2diii**	60 dm^3			
	1d	48 g		**3ci**	54 g			
	1ei	4.8 g		**3cii**	10.8 g			
	1eii	96 g		**3ciii**	9 g			
	1eiii	48 g		**3civ**	28.5 g			

ACKNOWLEDGEMENTS

The authors and publishers would like to thank the following for their permission to reproduce their photographs:

Part-title photograph for 'Atoms' Mike McNamee/Science Photo Library; 1.1 Dr Jeremy Burgess/Science Photo Library; 2.3 NASA/Science Photo Library; 3.2 Andrew McClenaghan/ Science Photo Library; 4.4 Telstar Transport Engineering Ltd; page 25 courtesy Anglo-Australian Observatory; 10.5 F. Walther/Zefa; 11.1 David Lean/Science Photo Library; page 35 Dr Jeremy Burgess/Science Photo Library; 18.4 Zefa/Eigen/Zefa; 18.5 Philippe Plailly/Science Photo Library; part-title photograph for 'Air and Water' Mark Boulton/ICCE; 22.1 NASA; 23.2 Planet Earth Pictures/P.J. Palmer; 23.3 Claude Nuridsany and Maria Perennou/Science Photo Library; 25.2 CNRI/Science Photo Library; 25.3 courtesy of Sodastream Ltd; 25.4 James Stevenson/Science Photo Library; 26.3 Justin Munro; 26.5 Greenpeace/ Eriksen; part-title photograph for 'Chemical Reactions' Lawrence Migdale/Science Photo Library; 31.2 Andrew Lambert; 31.3 Farmers Weekly Picture Library; 32.1 Frank Lane Picture Agency; 32.2 reproduced by kind permission of the Dean and Chapter of Lincoln; page 85 Q10 National Power PLC, Research & Technology; 33.2 R. Bond/Zefa; 33.3 Manfred Kage/Science Photo Library; ·33.4 Chuck Fell/Colorific!; 34.2 Adam Hart-Davis/Science Photo Library; 36.1 Valor Heating; 36.2 Bob Thomas Sports Photography; 36.3 Pentaprism; 36.4 James Holmes, Hays Chemicals/Science Photo Library; 39.3 (top) WWF/UK/J. Plant; 39.3 (bottom) WWF/UK/Mike Corley; 40.2 ICI Chemicals & Polymers Ltd; 39.4 Malcolm Fielding, Johnson Matthey PLC/Science Photo Library; part-title photograph for 'The Earth' Simon Fraser/Science Photo Library; ·cience Features Picture ·.2 Martin Land/Science· ·ary; 44.4 Steffen

Hauser/Oxford Scientific Films; 45.1 and 45.4 Geoscience Features Picture Library; 45.6 Rex Features/SIPA-Press; 46.2 W. Higgs/Geoscience Features Picture Library; 46.4 Stewart Lowther/Science Photo Library; 46.5 Rex Features/SIPA-Press; 46.6 The Hutchinson Library/Robert Francis; 47.1 Dr B. Booth/Geoscience Features Picture Library; 47.2 Geoscience Features Picture Library; 47.4 Niall Clutton/Arcaid; 47.7 Geoscience Features Picture Library; 47.9 George Holton/Science Photo Library; 48.2 a–c Geoscience Features Picture Library; 48.3 Dr B. Booth/Geoscience Features Picture Library; 48.4 Richard Bryant/Arcaid; 48.5 C. Carvalho/Frank Lane Picture Agency; 48.6 Landform Slides; 48.7 S. Stammers/Science Photo Library; 49.2 Mark Fiennes/Arcaid; 49.3 Jeremy Cocayne/Arcaid; 50.4 Travel Photo International; 51.2 Frank Lane Picture Agency/Roger Wilmhurst; 51.3 Landform Slides; 51.7 David Parker/Science Photo Library; 52.4 S. Stammers/Science Photo Library; 52.5 Landform Slides; 53.1 Frank Lane Picture Agency/C. Carvalho; 53.3 Holt Studios/Nigel Cattlin; 53.4 S. Stammers/Science Photo Library; 54.2 Landform Slides; 54.3 Siegfried Kerscher/Silvestris/Frank Lane Picture Agency; 54.4 Landform Slides; 54.5 Norbert Rosing/Silvestris/Frank Lane Picture Agency; 54.6 Leo Batten/Frank Lane Picture Agency; 55.2 Steve Percival/Science Photo Library; 55.4 Holt Studios/Mary Cherry; 55.5 Landform Slides; 56.4 Simon Fraser/Science Photo Library; 56.5 Michael Marten/Science Photo Library; 57.1 Krafft/Explorer/Science Photo Library; 57.2 Sheridan/Ancient Art & Architecture Collection; 57.3 George Bernard/Science Photo Library; part-title photograph for 'Resources' Zefa/Slatter; 58.1 Holt Studios/Nigel Cattlin; 58.2 The Hutchinson Library/Richard House; 58.3 Shell International; 58.4 The Environmental Picture Library/Philip Carr; 59.2 Dr B. Booth/Geoscience Features Picture

Library; 59.3 Zefa/Slatter; 60.2 British Coal; 60.3 Shell International; 66.3 Mike Devlin/Science Photo Library; 66.6 Jerry Mason/Science Photo Library; 66.7 Dr Morley Reader/Science Photo Library; 66.8 ROSPA; 67.1 Martin Bond/Science Photo Library; 67.2 courtesy British Cement Association; 67.3 and 67.4 courtesy of Pilkington plc; 68.1, 68.2 and 68.3 C.M. Dixon; 68.5 Hans-Frieder and Astrid Michler/Science Photo Library; 68.6 Travel Photo International/Rita Bailey; 70.2 Thermit Welding (GB) Limited; 71.3 Quadrant Picture Library; 72.2 and 72.3 Geoscience Features Picture Library; 72.4 Photo Library International; 74.1, 74.3, 74.4 and 74.5 Alcan Aluminium Ltd; 75.1 Geoscience Features Picture Library; 75.4 The Trustees of the Imperial War Museum, London; 76.1 Shell International; 77.1 Leslie Garland Picture Library; 77.2 IMI Range Limited; 77.4 Dr B.D. Dunn; 78.1 Kirk Precision Ltd; 78.2 U.S. Department of Energy/Science Photo Library; 78.3 Alcan Aluminium Ltd; part-title photograph for 'Periodic Table' IBM UK Ltd; 80.4 Tessa Traeger Library; 81.1 Andrew Lambert; 82.2 NOVOSTI/Science Photo Library; 82.3 Bill Brooks/Telegraph Colour Library; 85.3 Holt Studios/Nigel Cattlin; 88.1 Andrew Lambert.

Figures 47.5 and 52.2 are © Geoff Jones. All other photographs were taken and supplied by Graham Portlock, Pentaprism. All artwork is by Geoff Jones with the exception of Figure 71.2.

The authors and publishers would also like to thank the Nature Conservancy Council for supplying data on the erosion of an SSSI used in Q2 at the end of the section on 'The Earth'.

Every effort has been made to reach copyright holders; the publishers would like to hear from anyone whose rights have been unwittingly infringed.